THE YEAST ROLE IN MEDICAL APPLICATIONS

Edited by **Waleed Mohamed Hussain Abdulkhair**

The Yeast Role in Medical Applications
http://dx.doi.org/10.5772/intechopen.69408
Edited by Waleed Mohamed Hussain Abdulkhair

Contributors

Hongde Liu, Weiheng Ma, Jiahao Xie, Huamei Li, Kun Luo, Donghui Luo, Lei Liu, xiao sun, Francisco Navarro, Michael Altmann, Alina Kunicka-Styczyńska, Marta Targalska, Takeo Usui, Takumi Chinen, Yoshio Hayashi, Keisuke Hamada, Akihiro Taguchi, Yukihiro Asami, Kazuro Shiomi, Yoshiharu Inoue, Wataru Nomura, Jorge Gracida, Aline López Vargas, Aldo Amaro, Gurusamy Kutralam-Muniasamy, Fermin Pérez Guevara

Published by InTech
Janeza Trdine 9, 51000 Rijeka, Croatia

Publishing Process Manager Iva Simcic
Technical Editor SPi Global
Cover InTech Design team

First published January, 2018
Printed in Croatia
Legal deposit, Croatia: National and University Library in Zagreb

Additional hard copies can be obtained from orders@intechopen.com

The Yeast Role in Medical Applications, Edited by Waleed Mohamed Hussain Abdulkhair
p. cm.
Print ISBN 978-953-51-3734-4
Online ISBN 978-953-51-3735-1

For EU product safety concerns:
IN TECH d.o.o., Prolaz Marije Krucifikse Kozulić 3, 51000 Rijeka, Croatia,
info@intechopen.com or visit our website at intechopen.com.

Meet the editor

Waleed Mohamed Hussain Abdulkhair is a researcher of microbiology at the National Organization for Drug Control and Research, Egypt. He obtained his BSc degree in 2000 and was awarded his MSc degree in microbiology in 2007 and PhD degree in microbiology in 2011 from Al-Azhar University, Egypt. He worked as a lecturer of microbiology and botany at Teacher's College, King Saud University, Riyadh, Saudi Arabia from 2009 to 2014. His major research interests are medical microbiology, microbial fermentation biotechnology, and biological control. He has attended a lot of specific conferences in those fields and published many scientific papers that support these fields. He has participated as an author in two InTechOpen books entitled "Enzymes and Phytohormones from *Micromonospora*. In: Actinobacteria: Basics and Biotechnological Applications, (2015)" and "Plant Pathogens. In: Plant Growth, (2016)."

Contents

Preface

Yeast is a unicellular-eukaryotic microorganism belonging to the kingdom mycota. This genus has more than 2000 species, which represent approximately 1.5% of the total identified fungi. Although most yeasts are unicellular, some of them are multicellular due to pseudohyphae formation. The main reproduction way is mitosis, while budding and fission take place. Yeast was discovered thousands of years ago, where it ferments the carbohydrates and converts them into carbon dioxide and alcohol. The former is usually used in baking, while the latter is used in many industries including alcoholic beverages. Yeast was considerably used in a lot of biological researches including molecular biological ones due to its simple structure. Yeast such as *Saccharomyces cerevisiae* was known as a useful microorganism, but there are other species such as *Candida albicans* that are pathogenic to humans and animals. Yeast is described as a multimachinery microorganism due to its high metabolic production. For example, yeast generates electricity in the microbial fuel cells and produces ethanol under anaerobic conditions for the biofuel industry. Moreover, yeast can produce specific proteins called "killer toxins," which are used in the treatment of candidiasis. Killer toxins are also considered self-defense mechanism against other microorganisms. There are different habitats for yeast including soil, insects, sea, fruit surface, plant exudates, flowers' nectaries, ants, and gut of mammals as normal flora. Some genera such as *Cryptococcus neoformans* and *Cryptococcus gattii* are opportunistic pathogens, which cause cryptococcosis. *Candida glabrata* is a pathogenic yeast that causes candidiasis in the urogenital tract and candidemia.

Yeasts are significantly used in the fermentation biotechnological field, such as ethanol, bread, wine, and xylitol industries.*Yarrowia lipolytica* degrades oils, TNT, alkanes, fatty acids, and fats and has the ability to tolerate high salinity and heavy metals. *Saccharomyces cerevisiae* has the ability to bioremediate the toxic industrial effluents including arsenic and bronze statues. On the other hand, S. cerevisiae can be used as a nutritional supplement because it has high content of proteins, vitamins, and minerals and low content of fats and sodium ions. *Saccharomyces boulardii* is mainly used to maintain the natural flora of the gastrointestinal tract, decreases the acute diarrhea, and diminishes the harmful effect of *Clostridium difficile*. Moreover, yeasts are used in the bioremediation of the wastes. Although a lot of yeasts are used as food supplements, some of them are the main cause of food spoilage such as *Zygosaccharomyces* spp. and *Brettanomyces bruxellensis*.

Saccharomyces cerevisiae was used as a best eukaryotic model for determining the molecular biological features, physiological properties, and cell biology. For example, specific human proteins such as cell cycle proteins, signaling proteins, and protein-processing enzymes were discovered first in the yeast. The genomic structure of *Saccharomyces cerevisiae* genome sequence was first determined where it was composed of 12 million base pairs. In 2012, the full genome sequence of *Schizosaccharomyces pombe* was determined where it was composed of 13.8 million base pairs. Genetic engineering was used to improve the genetics of yeasts in particular S. cerevisiae for producing of pharmaceutical products such as insulin, vaccines, and human serum albumin and also for producing different chemicals such as phenolics, isoprenoids, alkaloids, and polyketides.

This book is a scientific work that aims to illustrate the vital role of yeast in biotechnology approach, in particular the medical field. I would like to thank the authors who have done a commendable job and worked hard in writing the chapters. Also, thanks to all contributors who have done hard work to bring this book to fruition. I wish that this book will be useful for scientists and researchers. I wish to appreciate and thank the efforts made by InTechOpen in publishing this book.

Waleed Mohamed Hussain Abdulkhair
National Organization for Drug Control and Research,
Egypt

Section 1

Benefits of the Yeast

Chapter 1

The Benefits of *Saccharomyces boulardii*

Michael Altmann

Additional information is available at the end of the chapter

http://dx.doi.org/10.5772/intechopen.70591

Abstract

A *Saccharomyces boulardii* strain, which does not carry any auxotrophic markers, was transformed with knockout constructs for the genes HIS3 and ADE2 using the dominant antibiotic marker genes encoding for kanamycin/G418- and nourseothricin/NAT^R resistance. Thereby, homozygous derivatives that were histidine or adenine deficient were obtained. Histidine prototrophy was easily reconstituted by transforming his-defective diploid derivatives with yeast plasmids carrying the HIS3 gene. Despite different attempts, for example, by creating a rme1::KANX rme1::NATR double-deleted *S. boulardii* yeast strain (RME1 encodes for Regulator of Meiosis), no visible sporulation to obtain haploid derivatives could be obtained. Besides, no filamentation properties of *S. boulardii* were observed. As previously mentioned, this yeast strain was confirmed to thrive at 37°C, a temperature disliked by some but not all *S. cerevisiae* strains used in the laboratory. *S. boulardii* is a diploid derivative of *S. cerevisiae* that does not sporulates and survives at temperatures as those found in the human gut. It can be easily manipulated by using conventional yeast methods to introduce auxotrophic markers and obtain heterozygous diploid knockout derivatives that can be transformed with yeast plasmids following conventional yeast protocols, thereby it could be even suited for biochemical and genetic research purposes.

Keywords: *Saccharomyces cerevisiae*, *Saccharomyces boulardii*, probiotic, genetic properties, yeast transformation

1. Introduction

The French microbiologist Henry Boulard isolated in 1923 a yeast strain (later named after him) after observing natives in Indochina affected by digestive disorders to chew litchi and mangosteen skins. It was said (but never proven) that those people could even protect themselves thereby against outbreaks of cholera.

Ever since, there has been an increasing body of medical reports addressing the beneficial properties of *Saccharomyces boulardii* as a probiotic to treat cases of diarrhea, reconstituting the gut flora after antibiotic treatment and even in the treatment of patients suffering from *Helicobacter pylori* that can cause stomach ulcers (for a recent review, see [1]). High doses of lyophilized living cells (2–5 × 10^9) are administered in pills that dissolve in the gastrointestinal tract. It is assumed that *S. boulardii* cells do not establish in the gastrointestinal tract and are secreted with the feces. Therefore, several doses are recommended to be taken during several days. It has not been clearly shown if on their passage yeast cells interact with other microorganisms in the gastrointestinal tract. Despite several clinical studies indicating beneficial effects of *S. boulardii,* there have been also reports identifying it as the cause of fungemia in hospital patients (adults and children) that had received doses of this yeast to treat gastrointestinal disorders [2]. Though the beneficial effects of the baker's yeast *S. cerevisiae* used for fermentation are unquestioned, certain wild *S. cerevisiae* isolates can have negative effects especially in people with a compromised immunosystem after undergoing surgery. Non-domesticated yeast strains as those isolated in hospitals can have invasive properties very different from those observed for domesticated yeast strains used in bakeries and breweries.

Yeast strains used in the laboratory such as S288C are mostly derivatives of industrial yeast strains used for ages in breweries [3]. More recently, derivatives of diploid strain Σ1278b, which shows filamentous properties related to non-domesticated yeast strains [4], are used in molecular biology research. Surprisingly, the sequence identity of both yeast strains is only 46% [5] indicating considerable genetic variability due to adaptation to differing milieus and to human domestication of this eukaryotic species [6].

Thanks to molecular genetic techniques, *S. boulardii* considered originally as a yeast species by itself [7] has been proven in recent years to be a variant of *S. cerevisiae* [8]. In this chapter, a diploid yeast strain similar to those strains used in most research laboratories, amenable to genetic manipulation when using conventional yeast protocols is shown.

2. Results

2.1. Growth properties of *S. boulardii*

To further characterize this yeast, its growth properties at different temperatures are compared with other diploid yeast strains (all yeast strains used are summarized in **Table 1**). *S. boulardii* (I will keep this name in the text even though it is a *S. cerevisiae* strain) grows well on rich medium at 30°C as well as at 37°C but not at 40°C (**Figure 1A**). Growth at 37°C—though not ideal—is not uncommon to yeast strains such as the diploid strain BY4743 (a derivative of S288C) but as opposed to diploid strain RH2585/2586 (an Σ1278b derivative) which hardly grows at 37°C (**Figure 1A**). In that sense, growth at 37°C is not a particular and unique property of *S. boulardii*. It probably rather reflects its accommodation to hot climates such as those found often in Indochina.

Name and properties of diploid yeast strain	Auxotrophic properties/antibiotic resistances
BY4743	Requires histidine, leucine, methionine, and uracil
RH2585/2586	Requires histidine and uracil
RH2585/2586 Δflo8::kanX Δflo8::NAT^R	No requirements; G418- and NAT-resistant
S. boulardii	None
S. boulardii Δflo8::kanX Δflo8::NAT^R	No requirements; G418- and NAT-resistant
S. boulardii Δrem1::kanX Δrem1::NAT^R	No requirements; G418- and NAT-resistant
S. boulardii Δhis3::kanX Δhis3::NAT^R	Requires histidine; G418- and NAT-resistant
S. boulardii Δade2::kanX Δade2::NAT^R	Requires adenine; G418- and NAT-resistant

G418/geneticin and NAT/nourseothricin are selective antibiotics for yeast strains.

Table 1. Diploid yeast strains used in this work.

A further investigated property is *S. boulardii's* potential capability to form filaments. For this purpose, it was grown on SLAD plates which carry only limiting concentrations of ammonium sulfate (50 μM, about 1000× less than conventional SD minimal medium). As observed under the microscope, the diploid strain RH2585/2586 clearly shows filamentous properties under such ammonium-limiting conditions (**Figure 1B**) [9]. Deletion of the FLO8 gene encoding a transcriptional factor required for filamentation and adhesion completely abolished its filamentous properties. Flo8 is required to express, for example, Flo11, a cell-surface glycoprotein [10]. As opposed to RH2585/2586, *S. boulardii* hardly showed any filamentation properties. Interestingly, sequencing of the PCR-amplified FLO8 gene of *S. boulardii* indicated that it does not carry a premature stop codon (not shown) found in non-filamentous yeast strains as those derived from S288C [10]. So, other up- or downstream genes required for filamentation are likely dysfunctional in *S. boulardii*. Its lack of filamentation probably explains its low toxicity and its lack of establishment capacity in the gastrointestinal tract that otherwise could make it more persistent and thereby more problematic for medical applications.

Another interesting issue was to induce meiosis and sporulation in *S. boulardii* in order to obtain haploid progeny. For this purpose, diploid cells were incubated in a liquid medium with limiting nitrogen and very high potassium acetate concentration as a (poor) carbon source [11]. Despite several attempts, no tetrad formation was observed. In that respect, *S. boulardii* shows similar properties as RH2585/2586 (the filamentous diploid used in this work) that does not form tetrads upon treatment under the described conditions. In order to induce meiosis and sporulation, a double knockout of RME1 (Regulator of Meiosis 1) in *S. boulardii* was produced. RME1 is a negative regulator of meiosis that prevents the expression of meiosis-required proteins such as IME1 (Inducer of Meiosis 1) and promotes mitosis [12]. The *S. boulardii* ΔΔrme1 derivative did not show any phenotypical differences to the parental wild-type strain in terms of growth temperature (**Figure 1A**). Unfortunately, no tetrads were obtained from this knockout strain either. I conclude that *S. boulardii* does not undergo meiosis and haploid tetrad formation at least under laboratory conditions used here.

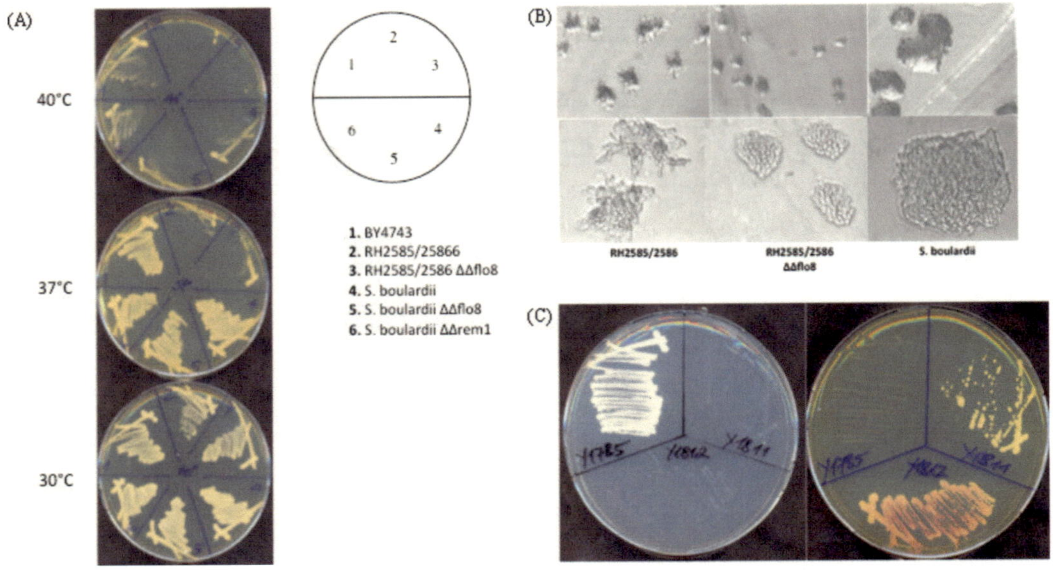

Figure 1. Properties of *S. boulardii* and other diploid yeast strains (all listed in **Table 1**). (A) Growth comparison of yeast strains at different temperatures. Indicated yeast strains were spread on YPD plates and incubated at indicated temperatures (30, 37, and 40°C) for 2 days; (B) filamentation properties of different yeast strains. Strains RH2585/2586, RH2585/2586 Δflo8::kanX Δflo8::NAT^R, and *S. boulardii* were incubated on SLAD plates (50 μM ammonium sulfate) for 1 day at 30°C and individual growth colony visualized under the microscope (upper panel: 20× magnification; lower panel: 100× magnification); (C) growth of *S. boulardii* on different media. *S. boulardii* (#1785) and derivatives ΔΔhis3 (#1811) and ΔΔade2 (#1812) were grown for 2 days at 30°C on minimal SD medium (left plate) or on YPD + G418 and nourseothricin (both at final 100 μg/ml) (right plate). Denote the characteristic pinkish color of strain #1812 due to the deletion of both ADE2 gene copies.

As shown in **Figure 1C**, *S. boulardii* does not carry any auxotrophic markers as it grows well on minimal medium (SD) devoid of amino acids or nucleic acid components such as adenine or uracil. Auxotrophic marker genes could be easily obtained by deleting the ADE2 (adenine biosynthesis) or HIS3 (histidine biosynthesis) genes (**Figure 1C**). These deletions were obtained by introducing dominant auxotrophic marker genes that provide resistance to the antibiotics kanamycin/G418 or nourseothricin. Deletion of a single gene copy of ADE2 or HIS3 still allowed for growth on minimal medium plates (not shown) indicating the clear diploid character of this yeast. Only double deletion of both HIS3- or ADE2-gene copies (which made it resistant against both kanamycin/G418 and nourseothricin; **Figure 1C**) made this yeast strain auxotrophic for histidine or for adenine. Newly gained histidine auxotrophy was used in a subsequent experiment to transform it with yeast plasmids carrying HIS3 as a selectable marker gene (see subsequent text).

2.2. Transformation of *S. boulardii* with conventional yeast plasmids

As an auxotrophic histidine-deficient yeast strain was now available, I decided to transform it with conventional yeast plasmids that complement for the lack of HIS3. For this purpose, *S. boulardii* ΔΔhis3 (#1811) was grown in minimal SD medium supplemented with histidine. Cells were made competent by treating them with Li-acetate following a well-established yeast transformation protocol [13].

After 2–3 days of incubation at 30°C, his$^+$-transformants were nicely observed (**Figure 2**; left panel), indicating that a simple yeast transformation protocol was sufficient to transform this yeast strain and to recover its prototrophy. Plasmids used for transformation (p301HIS3 GAL-p20-HA from *S. cerevisiae* and *Candida albicans*) allow for the expression of the protein p20 (a modulator of the activity of eIF4E, the cap-binding protein; see subsequent text) when growing cells in a medium containing galactose. As shown on a Western Blot obtained from yeast extracts, p20 from different sources was expressed in galactose but not in glucose-containing media (**Figure 2**; right panel). This confirms that, in *S. boulardii*, the GAL1/10 promoter is regulated in an identical manner as in conventional yeast strains used in the laboratory [14].

2.3. Sequencing and comparison of *S. boulardii* p20 gene

p20, a small acidic protein of 161 amino acids, is encoded by the non-essential gene CAF20 which only exists in a variety of yeast species (such as *S. glabrata, Kluyveromyces lactis, C. albicans, S. cerevisiae*). Its function is related to regulating the activity of the cap-binding protein eIF4E during translation in a yet not-well-understood manner [15]. A sequence alignment of different yeast species (see **Figure 3**; upper panel) shows a clear homology but not identity of the corresponding p20 proteins. Especially conserved are peptide motifs at the amino terminus (which are required for binding to eIF4E; the canonical motif YxxxxLL/I/F highlighted) and at the carboxy terminus (where precise function has still to be determined).

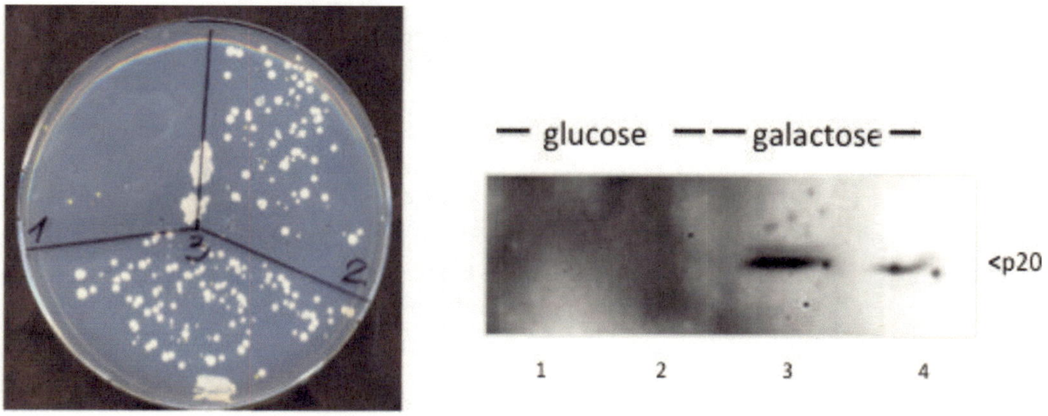

Figure 2. Transformation of *S. boulardii* ΔΔhis3 and expression of p20. (Left panel) Transformation of *S. boulardii* ΔΔhis3. *S. boulardii* ΔΔhis3 was transformed with plasmid p301HIS3 p20 (*S. cerevisiae*)-HA (segment 2) or with plasmid p301HIS3 p20 (*C. albicans*)-HA (segment 3) and grown on minimal medium plates (without histidine) for 3 days at 27°C. As a negative control, no plasmid DNA was added (segment 1); (right panel) expression of HA-tagged p20 after induction with galactose. *S. boulardii* ΔΔhis3 extracts from cells transformed with p301HIS3 p20 (*S. cerevisiae*)-HA (lanes 1 and 3) or transformed with p301HIS3 p20 (*C. albicans*)-HA (lanes 2 and 4) grown on medium containing 2% glucose or 2% galactose as indicated are shown. Individual colonies from segments 2 and 3 were picked and grown in liquid SD medium for 24 h. Subsequently, half of the cells were collected, washed (2× with water), resuspended in SGal (minimal medium with 2% galactose), and incubated for 24 h at 27°C. Collected cells were boiled in 2× SDS-sample buffer, proteins separated on a 15% SDS-PAGE gel. Separated proteins were blotted onto nitrocellulose, subsequently incubated with monoclonal antibodies against the HA-tag (1:2000 Dilution in 2% skim milk). For visualization of the Western Blot, WesternBright ECL kit (advansta) was used.

```
sp|Q5AQ12|CAF20_CANAL       MAKYTEEQLLELKSEAHTPKPEILDAFNKLIEEVKESIEQHQQHQRKWHNGDTYIDEHGH
tr|Q6CU66|CAF20_KLUYV       MIRYTEEELLQLRPTEPVKPNFDVDEFNAIIEKVKEIQEAHEEEFSH-------------
sp|Q6FKJ9|CAF20_CANGA       MIKYSIDELIQLKPSETLKVSFDHVEFRNIIEKVVELQKLKEEEFHSHHG----------
sp|P12962|CAF20_YEAST       MIKYTIDELFQLKPSLTLEVNFDAVEFRAIIEKVKQLQHLKEEEFNSHHVG---------
                            * :*: ::*::*:             *. :**:* :  . :::.

sp|Q5AQ12|CAF20_CANAL       ERSYHHINRRRQSKGASGVPRPNLR-KKSEPVVDEDGWATLSKPKKGSFAEGDAIEER--
tr|Q6CU66|CAF20_KLUYV       -------FRRRSSHH-HA--KPKFKHLKPKITTDEEGWSTLETAPAVRRKSVA-------
sp|Q6FKJ9|CAF20_CANGA       -------NRRRSSHH-HM--KPKIKHNKPKVKTDADGWSTLEPATAGHEEESSSSAT-PA
sp|P12962|CAF20_YEAST       -----HFGRRRSSHH-HG--RPKIKHNKPKVTTDSDGWCTFEAKKKGSGEDDEEETETTP
                                    ***.*:      :*:::  * :  .*  :**.*:.        .

sp|Q5AQ12|CAF20_CANAL       -IKFRETNNSGAGIKARPNNKNLGSSKAVDPREIASDKQTKAFNAFAALGDEDDDDEDDE
tr|Q6CU66|CAF20_KLUYV       EEEEPTIVIAQETLKVKPN-KHISSSRPADARDIVADKPSKAFNAFAALESDEEEEQE--
sp|Q6FKJ9|CAF20_CANGA       AAATTKTGAPQETIRVKPNNKNISSSRPADNSDIIADKQTHGFNAFAALEDEEDE-----
sp|P12962|CAF20_YEAST       TSTVPVATIAQETLKVKPNNKNISSNRPADTRDIVADKPILGFNAFAALESEDEDDEA--
                                         ::.:** *::.*.: .*  :* :**   .*******  .::::


Sboulardii                  MIKYTIDELFQLKPSVTLEVNFDAVEFRAIIEKVKQLQHLKEEEFNSHHVGHFGRRRSSH
sp|P12962|CAF20_YEAST       MIKYTIDELFQLKPSLTLEVNFDAVEFRAIIEKVKQLQHLKEEEFNSHHVGHFGRRRSSH
                            ***************:********************************************

Sboulardii                  HHGRPKIKHNKPKVTTDSDGWCTFEAKKKGSGEDDEEETETTPTSTVPVATIAQETLKVK
sp|P12962|CAF20_YEAST       HHGRPKIKHNKPKVTTDSDGWCTFEAKKKGSGEDDEEETETTPTSTVPVATIAQETLKVK
                            ************************************************************

Sboulardii                  PNNKNISSNRPADTRDIVADKPILGFNAFAALESEDEDDEA
sp|P12962|CAF20_YEAST       PNNKNISSNRPADTRDIVADKPILGFNAFAALESEDEDDEA
                            *****************************************
```

Figure 3. Sequence comparison of p20 from different yeast species. Multiple sequence alignment of p20 from different yeast sources (CANGA, *Candida glabrata*; KLUYV, *Kluyveromyces lactis*; CANAL, *Candida albicans*; YEAST, *Saccharomyces cerevisiae*) was done with the help of Clustal Omega (http://www.ebi.ac.uk/Tools/msa/clustalo/). Highlighted amino acids; see text. For sequencing of *S. boulardii* p20 gene, the complete ORF and adjacent sequences were PCR-amplified with oligonucleotides used for *S. cerevisiae* p20 gene amplification and the obtained PCR product was sequenced in both directions (shown is the part encoding the p20 open reading frame of *S. boulardii*).

The CAF20 gene from *S. boulardii* was PCR-amplified using genomic DNA and oligonucleotides hybridizing at the 5′ and 3′ region of the *S. cerevisiae* CAF20 gene. Subsequent sequencing showed that CAF20 from *S. cerevisiae* and *S. boulardii* is nearly identical. The only difference detected is a conserved amino acid substitution (leucine to valine; highlighted) at position 16 (**Figure 3**; lower panel). Among those yeast species that carry the p20 gene conservation varies between 30 and 90% (not shown). The almost identity of both sequences shown here clearly confirms that *S. boulardii* is a variant of the species *S. cerevisiae*.

3. Conclusions

In this work, I present data indicating that *S. boulardii* is a diploid *S. cerevisiae* strain. It thrives well under laboratory conditions at different temperatures (up to 37°C) which is not unusual for different laboratory yeast strains. *S. boulardii* does not show filamentous properties even though its FLO8 gene does not carry the typical premature stop codon identified in many laboratory (and industrial) yeast strains. *S. boulardii* does not undergo meiosis or form haploid progeny when incubated in sporulation-inducing media. *S. boulardii* does not carry any identifiable auxotrophic gene markers. It can be easily manipulated to obtain knockout derivatives by inserting genes conferring antibiotic resistance and obtain thereby auxotrophic progeny. Additionally, *S. boulardii* can be easily transformed with conventional yeast plasmids allowing also for the expression of proteins regulated by the galactose-inducible GAL 1/10 yeast promoter.

In accordance with those properties, it is probably not detrimental for human health (at least not for immunocompetent individuals) as it will not easily establish in the gut or penetrate the intestinal blood barrier. All this does not mean that it has beneficial physiological properties and I would like to ask the question: is this not just a further conventional yeast strain?

Acknowledgements

S. boulardii was obtained from I. M. Castro (Belo Horizonte/Brazil) [16]. I would like to thank Daniela Ross-Kaschitza from my laboratory for carefully reading this manuscript. This work was supported by the Swiss National Foundation grant 31003A_146722/1.

Author details

Michael Altmann

Address all correspondence to: michael.altmann@ibmm.unibe.ch

Institut für Biochemie und Molekulare Medizin (IBMM), Universität Bern, Bern, Switzerland

References

[1] McFarland LV. Systematic review and meta-analysis of saccharomyces boulardii in adult patients. World Journal of Gastroenterology. 2010;**16**:2202-2222

[2] Muñoz P, Bouza E, Cuenca-Estrella M, Eiros JM, Perez MJ, Sanchez-Somolinos M, Rincon C, Hortal J, Pelaez T. *Saccharomyces cerevisiae* fungemia: An emerging infectious disease. Clinical Infectious Diseases. 2005;**40**:1625-1634. DOI: 10.1086/429916

[3] Mortimer RK, Johnston JR. Genealogy of principal strains of the yeast genetic stock center. Genetics. 1986;**113**:35-43

[4] Gimeno CJ, Ljungdahl PO, Styles CA, Fink GR. Unipolar cell divisions in the yeast *S. cerevisiae* lead to filamentous growth: Regulation by starvation and RAS. Cell. 1992;**68**:1077-1090

[5] Dowell RD, Ryan O, Jansen A, Cheung D, Agarwala S, Danford T, Bernstein DA, Rolfe PA, Heisler LE, Chin B, Nislow C, Giaever G, Phillips PC, Fink GR, Gifford DK, Boone C. Genotype to phenotype: A complex problem. Science. 2010;**328**:469. DOI: 10.1126/science.1189015

[6] Liti G, Carter DM, Moses AM, Warringer J, Parts L, James SA, Davey RP, Roberts IN, Burt A, Koufopanou V, Tsai IJ, Bergman CM, Bensasson D, O'Kelly MJ, van Oudenaarden A, Barton DB, Bailes E, Nguyen AN, Jones M, Quail MA, Goodhead I, Sims S, Smith F, Blomberg A, Durbin R, Louis EJ. Population genomics of domestic and wild yeasts. Nature. 2009;**458**:337-341. DOI: 10.1038/nature07743

[7] McFarland LV. *Saccharomyces boulardii* is not *Saccharomyces cerevisiae*. Clinical Infectious Diseases. 1996;**22**:200-201

[8] Douradinha B, Reis VC, Rogers MB, Torres FA, Evans JD, Marques ET, Jr. Novel insights in genetic transformation of the probiotic yeast *Saccharomyces boulardii*. Bioengineered. 2014;**5**:21-29. DOI: 10.4161/bioe.26271

[9] Ross D, Saxena M, Altmann M. eIF4E is an important determinant of adhesion and pseudohyphal growth of the yeast *S. cerevisiae*. PloS One. 2012;**7**:e50773. DOI: 10.1371/journal.pone.0050773

[10] Liu H, Styles CA, Fink GR. *Saccharomyces cerevisiae* S288C has a mutation in FLO8, a gene required for filamentous growth. Genetics. 1996;**144**:967-978

[11] Jakubowski H, Goldman E. Evidence for cooperation between cells during sporulation of the yeast *Saccharomyces cerevisiae*. Molecular and Cellular Biology. 1988;**8**:5166-5178

[12] Shimizu M, Li W, Covitz PA, Hara M, Shindo H, Mitchell AP. Genomic footprinting of the yeast zinc finger protein Rme1p and its roles in repression of the meiotic activator IME1. Nucleic Acids Research. 1998;**26**:2329-2336

[13] Gietz RD, Schiestl RH. Applications of high efficiency lithium acetate transformation of intact yeast cells using single-stranded nucleic acids as carrier. Yeast. 1991;**7**:253-263. DOI: 10.1002/yea.320070307

[14] Guarente L, Yocum RR, Gifford P. A GAL10-CYC1 hybrid yeast promoter identifies the GAL4 regulatory region as an upstream site. Proceedings of the National Academy of Sciences United States of America. 1982;**79**:7410-7414

[15] Altmann M, Schmitz N, Berset C, Trachsel H. A Novel inhibitor of cap-dependent translation initiation in yeast: p20 competes with eIF4G for binding to eIF4E. EMBO Journal. 1997;**16**:1114-1121. DOI: 10.1093/emboj/16.5.1114

[16] Fietto JL, Araujo RS, Valadao FN, Fietto LG, Brandao RL, Neves MJ, Gomes FC, Nicoli JR, Castro IM. Molecular and physiological comparisons between *Saccharomyces cerevisiae* and *Saccharomyces boulardii*. Canadian Journal of Microbiology. 2004;**50**:615-621. DOI: 10.1139/w04-050

Section 2

Resistant Yeasts for Multi-antibiotics

Chapter 2

Candida Biofilms: Environmental and Clinical Aspects

Marta Targalska and Alina Kunicka-Styczyńska

Additional information is available at the end of the chapter

http://dx.doi.org/10.5772/intechopen.70703

Abstract

Candida spp. strains are characterized by their ability to form a biofilm structure on biotic and abiotic surfaces, causing significant problems in many industrial branches and threatening human health. *Candida* biofilm is a heterogeneous, spatially well-organized structure consisting of planktonic and mycelial yeast forms which are interdependent in the *quorum sensing* system and surrounded by an extracellular polysaccharide substance. Biofilm-forming microorganisms are characterized by high invasiveness, the ability to cause dangerous and difficult to treat infections. Furthermore, the cells in the biofilm, compared to planktonic forms, show reduced sensitivity to chemical compounds with antifungal activity and increased survival under unfavorable environmental conditions. The chapter focuses on the emergence of antifungal resistance with the development of biofilms. The work presents the examples of antibiotic resistance of a variety of *Candida*, showing that a group of strains expressing intermediate sensitivity or resistance to the tested antibiotics include both clinical and food-borne isolates. Similarities in enzymatic and biochemical profiles of different origin isolates are discussed. A substantial heterogeneity within *Candida albicans* group is also underlined. Simultaneously, the incidents of biochemical profiles conformity of some clinical and food-borne isolates are presented, which may be a result of *Candida* transmission via food.

Keywords: *Candida albicans*, non-*albicans Candida*, *Candida* biofilm, drug resistance, food-borne *Candida*

1. Introduction

Unicellular forms of yeast are rarely found in nature as single, scattered cells, in the form of plankton but they are rather adsorpt at the solid-liquid, liquid-gas, or liquid-liquid interface. Generally, they form organized, settled structures taking the form of multicellular clusters forming biofilm. Biofilm, also called as the biological membrane, is a complex, multicellular, and multifunctional structure of one or more species of microorganisms, surrounded by a layer of organic and inorganic substances produced by these microorganisms adhering to both biotic and abiotic surfaces. The form of biofilm enhances the effectiveness of microbial protection against the adverse environmental factors, including antibiotics, reduces the effectiveness of host defense mechanisms, facilitates the acquisition of nutrients, creates the possibility of horizontal gene transfer by providing evolutionary and genetic diversity, and enables the transmission of information between microbial cells [1–3].

Biofilm is most commonly formed on solid surfaces staying in contact with water, living tissues, and liquid-air interface. This ubiquitous structure can be very useful but also dangerous being difficult to be removed. Biofilm plays a key role in a process of self-cleaning of surface-, ground-, and underground water. The biofilm's ability to create a biobarrier has been exploited in water treatment and to reduce a pollution of soil and ground waters. Biofilm also allows biological removal of pollutants from sewage [2]. Biofilm exists not only in the natural environment but also is industrially applied, for example, to catalyze complex chemical reactions. Natural microbiota of the body of a healthy person forms a biofilm modulating some physiological functions, for example, colonic biofilm [4]. Moreover, changing environmental conditions may transform a biofilm from a big friend into a fierce enemy. A good example is the biofilm of the gastrointestinal tract, which, in unfavorable conditions, can become a source of mortal danger. In public facilities such as hospitals, hotels, swimming pools, physiotherapeutic facilities, sanatoria, mass caterers, schools and kindergartens, homes, and enterprise of the cosmetic and food industries, biofilm structure allows saprophytic and pathogenic microorganisms to survive washing, cleaning, and disinfection processes. Biofilm formed in a water supply network poses a sanitary risk to the public. In addition, the pipes water network is subjected to microbiological corrosion. Most food processing plants are struggling with the problem of biofilm formation in water distribution systems, refrigeration systems, and heat exchangers. In the food industry, biofilm can colonize not only sewage systems, but also machine working surfaces and food products. Biofilm on work surfaces, even those made of stainless steel, glass, or Teflon, can lead to food contamination with spoilage microorganisms, including pathogenic ones. Contaminated products of both plant and animal origin can cause serious human illnesses as well as huge losses in the food industry [2]. Biofilm microorganisms are characterized by increased invasiveness and the ability to cause serious infections, even in hospital. Ability to create biofilm is one of the pathogenicity factors of the microorganism. Most often, infections caused by biofilm-building microorganisms are the result of the abiotic surfaces colonization and account about 65% of all infections [1]. Microorganisms inhabiting medical materials both biomaterials within the human body such as vascular and intraperitoneal catheters, artificial valves, prostheses, implants, lenses, stitches, and diagnostic devices such as endoscopes, fibroscopes, and laryngoscopes are also an important problem. Biofilm formation on these devices is the cause of serious infections and also leads to device damage [1, 2, 5–7]. Microorganisms that inhabit the human body also occur mainly in the form of biofilms. These biofilms are mostly composed of symbiotic microorganisms, but also opportunistic ones may occur, which in homeostasis disturbances lead to a development of serious infections. The situation is particularly dangerous, if the development of infection is accompanied by a dysfunction of the device colonized by biofilm.

2. Biofilm definition

Biofilm is defined as a well-organized, three-dimensional social structure surrounded by extracellular matrix and irreversibly bound to the surface, built by microorganisms with altered, with respect to planktonic form, genotype properties [5, 6, 8–10]. Biofilm enables microorganisms to survive in a changing and unfavorable environment, and therefore is the dominant form of their existence in the nature. It is characterized by structural heterogeneity, genetic diversity, complexity of interaction, and the presence of extracellular substances. It can be either mono- or multilayer, produced by one species or many different species. The biofilm structure depends on many factors such as hydrodynamic conditions, surface type, pH of the environment, microbial mobility, intercellular communication, nutrient content, exopolysaccharides, proteins, or oxygen. Colonization of various surfaces by microorganisms is possible due to their adhesive properties and extracellular polymeric substances (EPS) stabilizing the biofilm structure. Adjacent microorganisms, in a spatially organized structure, produce a common layer of polymeric substances called extracellular matrix, the complex compounds playing an important role in the formation and functioning of the biofilm. Most EPS polysaccharides are the organic compounds with long linear or branched molecules of 106 Da. The amount of polymers depends on the quantitative and qualitative composition of nutrients. The percentage of water in the biofilm matrix is up to 97%. Polymers ability to cyclical accumulation simultaneously with donation of water gives the matrix hydrogel features with exceptional viscoelastic properties [2, 11–14]. Matrix hydrogel nature effectively protects biofilm microorganisms from desiccation and provides the cells with protection against environmental stress factors such as UV radiation, temperature shifts, pH fluctuations, or toxic substances [2, 5, 7]. The matrix serves also as a communication system between biofilm cells, where chemical and physical signals are transmitted through a branched open channel system separating individual microcolonies. Thanks to the channel network, oxygen and nutrients are delivered through the channels and the excreted waste products are discharged. Cells in biofilms are present in various metabolic states. On the periphery of the biofilm, where the channel network system is more developed, the cells are large, metabolically active, and its reproducing increases the biofilm thickness. While, microorganisms located inside the biofilm are partially cut off from the water system, which results in their growth rate decreasing. They may also fall in an anabiosis with possible activation in a case of destruction of the outer cell layer, which, no matter how long the biofilm works, uses the features of young biofilm cells [2, 12, 13]. The biofilm cell has different characteristics than the planktonic cells. An important determinant of biofilm properties is *quorum sensing*, a specific communication system, strictly controlled by specific genes in response to the abundance of cells in the biofilm—the sense of the piston. The ability of cells to communicate makes the biofilm able to function in a way that resembles an almost one multicellular organism consisting of physiologically diverse subpopulations of microbial cells [2, 4, 5, 12, 13, 15].

3. Biofilm structure

The process of biofilm formation is multistage and depends on the properties of the microorganisms, the construction, and properties of the colonized materials or the host. There are four basic phases: (I) reversible adhesion, (II) irreversible adhesion, (III) biofilm maturation, and (IV) dispersion (**Figure 1**).

Figure 1. Biofilm formation phases: (I) reversible adhesion, (II) irreversible adhesion, (III) biofilm maturation, and (IV) dispersion (elaborated according to [2]).

Biofilm formation begins with the adhesion of free floating microorganisms to the biotic or abiotic surface. Reversible adhesion is the result of relatively weak physical interactions causing the first cells to attach to a solid surface such as gravitational interaction, electromagnetic surface charge, van der Waals forces, electrostatic, and hydro- and thermodynamic forces (Brownian motion). These forces play a crucial role when the distance between cells and the surface is relatively large. Biofilm is unstable and can easily be removed by both chemical and physical methods. When the cell distance from the surface is less than 1.5 nm, there is irreversible adhesion due to the formation of specific bonds. First microbial cells attached to the surface help attaching another one by the formation of hydrophobic, non-specific or specific hydrogen bonds, and pairs and ionic complexes (carbon-carbon covalent bonds) [2, 5, 12, 13]. An important place in the biofilm-building process is the interaction of specific receptors, adhesives, and ligands on the cell surface of the microorganism or the target host cell extracellular ligand. Initially, the surfaces are covered by a single layer of microbial cells. In the construction of the basic EPS matrix, which gives the biofilm a defined shape and structure, the increased synthesis and secretion of extracellular biopolymers is important. Biofilms expand by increasing the intensity of cell proliferation. While, glycocalyx, a shell composed of polysaccharide residues of glycolipids and glycoproteins, the components of the cell membrane, is produced up to the total surroundings of the microcolonies. At this stage, biofilm, in addition to living microorganisms, also includes dead cells, mineral substances, and organic compounds. These elements are joined by further microbial cells. Irreversible adhesion allows the formation of microcolonies and biofilm maturation [5, 12, 13]. Biofilm maturation is followed by the microorganisms' reproduction, their gradual differentiation and the activation or inhibition of expression of certain genes. Biofilm cells acquire features that are not expressed by planktonic cells and can transmit them to adjacent and progeny cells. When reaching the critical thickness of the biofilm membrane, cells migrate from peripheral parts of the mature biofilm to the surrounding environment and the process of colonization begins. Disconnecting cells from biofilm and its dispersion is an intentional separation resulting from a reaction to adverse environmental conditions. Biofilm adapts to environmental stresses and the detached cells begin the process of colonization of new surfaces [2, 3, 5, 8, 12, 13].

Both bacterial and fungal biofilms, in medicine and in industry, were first described in 1978 [7]. Since then, it has been the subject of numerous studies that aim to understand the molecular mechanisms of its origins and the role it plays in infections and drug resistance [5]. *Candida albicans* often occurs in the form of biofilm, which is the etiological factor of approximately 90% candidiasis. Among the clinical strains of the genus *Candida*, biofilm formation depends on the type of a strain [16], and *Candida albicans*, even of the same genotype, may differ in biofilm features [1].

4. *Candida* dimorphism and the biofilm formation

Compared to planktonic forms, biofilm cells lead settled lifestyles and have characteristic gene expression associated with the growth rate and synthesis of some of the adhesion and enzyme proteins. Fungal biofilms with cells differing phenotypically and functionally usually are of much more complex structure than the bacterial biofilm. Polymorphism is a characteristic feature of *Candida* yeast. The planktonic *Candida* are usually in a form of blastospores (budding cells), while the biofilm structure is formed by both blastospores and mycelial forms. During biofilm formation, morphological transformation takes place: from blastospores through the germ tubes to the filamentous forms (mycelium or pseudomycelium). The plasticity of *Candida* biofilm indicates that its cell composition may also change depending on the location and characteristics of the biomaterial surface. Individual cell types exhibit differences in antigenic structure and its adhesion and invasive properties, enzymatic activity, and phagocytosis resistance. Blastospores are responsible for the adhesion and spreading of the biofilm, initiating its production by adhering to biotic and abiotic surfaces and its colonization [5]. Adhesion is a signal that induces germination (morphogenesis) of blastospores and the formation of invasive forms, mycelium, or pseudomycelium, that enter epithelial or endothelial cells via endocytosis or active penetration and because of enzymatic activity they contribute to the destruction of colonized tissues. The presence of mycelial forms is not a prerequisite condition for the biofilm formation; however, it seems indispensable in the process of maturation. In the mature *Candida* biofilm, the inner layer is composed of blastospores, and the outer multilayer is mycelium and pseudomycelium. Biofilms created by *Candida* sp. yeast can reach a thickness of 25–450 μm [1, 5, 10, 11, 17]. There are three [6, 11] or four [1, 18] phases of *Candida* biofilm formation. Sometimes the last dispersion phase is included in the third phase, maturation of the biofilm.

5. *Candida* adhesion and the ability to the biofilm formation

The biofilm structure depends on the specific gene expression resulting from yeast contact with biotic or abiotic surface. *Candida albicans* yeast contact with a specific surface and activate the mitogen-activated protein kinase (MAPK) signaling cascade, which carries the extracellular contact signal to eukaryotic cells, then activating transcription factors and expressing a specific set of genes responsible for adherence. MAPK gene *mkc1* activity level is higher in cells growing on different surfaces than in planktonic cells. After contacting *Candida albicans* with a polystyrene surface, the transcriptional level of the gene coding for methionine and cysteine, and the *cdr1* and *mdr1* codes for the mechanism of active ejection of the drug by efflux are surprisingly increased. Some examples of *Candida* sp. adhesion to polystyrene are presented in **Figure 2**.

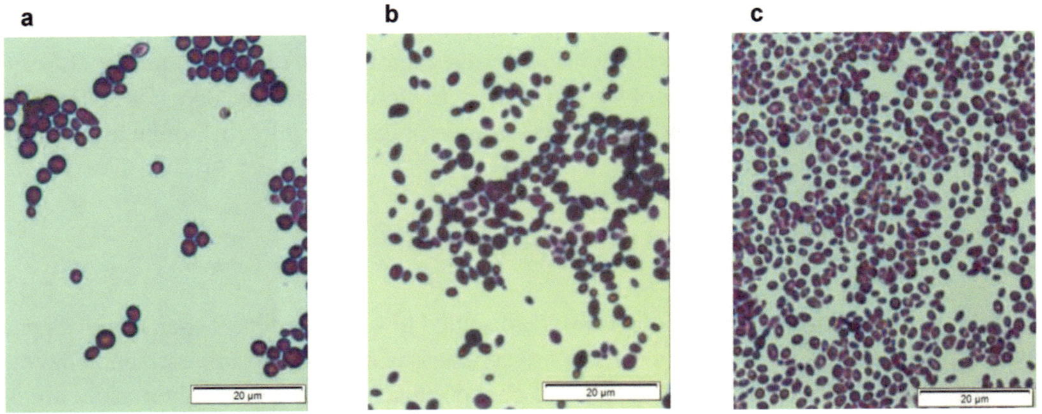

Figure 2. Adhesion to a polystyrene surface of environmental *Candida* sp. strains (a) *Candida albicans* cl/MP/12 clinical isolate; (b) *Candida parapsilosis* Fo/82/03 food-borne isolate; (c) *Candida albicans* cl/MP/08 clinical isolate (photographs by M. Maroszyńska).

Genes encoding sulfur amino acids are responsible for the amount of biofilm biomass produced, while the expression of the *cdr1* and *mdr1* genes is associated with the acquisition of drug resistance by the biofilm phenotype. What is more, the activity of adh1 alcohol dehydrogenase is higher in plankton cells than in biofilm cells, which influences the biofilm formation [1]. This enzyme is probably responsible for inhibition of biofilm formation and inactivation or mutation of the *adh1* gene results in stronger biofilm formation *in vitro* and *in vivo* [1, 5]. Adhesion to both biotic and abiotic surfaces related to recognition of host cell surface receptors, is a precondition for colonization, biofilm formation, initiation, and development of infection. Lack of adhesion prevents yeast cells from biofilm formation. Phenotypic variability and plasticity of cells in relation to changing environmental conditions allow for the settlement of new surfaces. Numerous *Candida albicans* gene products important for biofilm development have been identified [1, 19]. Adhesive genes can be activated by different environmental signals. The potential adhesives involved in biofilm formation are cell wall surface proteins. The input of surface proteins with GPI (glycosylphosphatidylinositol) module is decisive in the formation of fungal biofilm. Ability of *Candida albicans* to adhesion is an important virulence factor associated with glycoproteins encoded primarily by genes such as *hwp1* and *als*. Such an effect leads to changes in the expression of genes encoding cell wall proteins glycosylphosphatidylinositol dependent. Several key adhesins: als1, als2, als3, als4, als5, eap1, hwp1, hydrolases, lipases, phospholipases, and transcriptional factors bcr1 regulating protein expression are responsible for biofilm formation [1, 5, 19, 20]. Ywp1, mannoprotein with a GPI module, is both the regulator of adhesion and a marker of *Candida albicans* cells. Blastospores are characterized by weaker intercellular adherence than mycelial forms. In addition, they are probably equipped with mechanisms to regulate the activity of their own adhesins [1]. The ywp1 protein found on yeast cells, a known adherence regulator, can function as an antihistamine and inhibit adherence. The discovery of the ywp1 protein is a scientific proof for the presence of the cell diffusion phase in the biofilm's life cycle. The deletion of the ywp1 gene leads to the enhancement of the blastospor adherence to the various surfaces [1]. On the other hand, in the biofilm layer composed of pseudomycelium *Candida albicans* identified adhesins with GPI, HWP1, EAP1 module, and the ALS adhesin family. Some studies indicate that the genes *als* and *eap1* within the *Candida* genus show similarity to the genes regulating adhesion in *Saccharomyces cerevisiae* [1]. While, the *hwp1* gene, known as the gene coding for the main *Candida albicans* protein, is involved in many functions such as cell wall building, intracellular signaling, and the development of hyphae. In addition, it appears that it is involved in the adhesion of yeast to epithelial cells, which is so important in the initial colonization stage. There is also an evidence for the involvement of the *hwp1* gene in systemic candidiasis pathogenesis on mouse model *in vivo*. It has been shown that strains having the *hwp1* and *hwp1*-null heterologous genes showed, respectively, reduced and no virulence compared to control wild strains. Hwp1 is the first exposed adhesive, required for biofilm formation *in vivo*, which is not present on cells in the form of yeast and plays no role in the formation of microcolonies. However, it expresses during morphogenesis blastospores to pseudomycelium [1].

6. *Candida* germ tubes and the ability to the biofilm formation

Pseudomycelium is formed by a germ tube process and as a key component of the biofilm provides its integrity. Both morphological forms of blastospore and pseudomycelium are capable of a biofilm formation, but strains capable of growth only in the form of blastospores produce only residual biofilm. The transcription factor efg1 plays a key role in regulating the morphology and virulence of yeast *Candida albicans*. It was first identified as an inducer of the development of pseudomycelium in *Saccharomyces cerevisiae* and then as a necessary for the growth of mycelium *Candida albicans*. The consequence of deletion of this gene is a loss of ability to transition into mycelial forms in response to majority of stimulation factors, but may occur in hypoxia and abiotic conditions. The *efg1* gene fulfills many of the important functions in *Candida albicans* yeast cells, and most important is the virulence of vast infection models. Cells with *efg1* gene deletion do not attack the human epithelium. In addition, *efg1* is one of the key regulators of transition from the "white" form to the "opaque" form and essential to keep the default "white" phenotype. Moreover, unlike many other biofilm process regulators, the *efg1* gene is essential for biofilm development under hypoxia and oxygenation conditions. Even when the yeast cells have adhered to the abiotic surface it is necessary to produce resistance to antifungal agents. Efg1 is a part of a network of six transcription factors that regulate the expression of at least 1000 genes involved in the development of the *Candida albicans* biofilm [21]. Several thousands of intergenic regions bound by the transcriptional factor efg1, which binds to promotors at least 53 genes in *Candida albicans*, including many transcription factors have been identified. The binding of efg1 is closely related to the transition from the basic form of the yeast cell to the pseudomycelium. Mutants with the deletion of *efg1*, *cph1*, and *tec1* genes encoding transcription factors do not form pseudomycelium, and consequently have no ability to form mature biofilm structures. This indicates that transcriptional factors efg1, cph1, and tec1 play a key regulatory role in the formation of mature *Candida albicans* biofilm [1, 21]. Furthermore, the mutations of the genes *suv3*, *nup85*, *mds3*, and *kem1* inhibit the formation of pseudomycelium, which in turn promotes the formation of "immature" biofilm. In addition, mutants with the deletion of the *bcr1* gene produce pseudomycelium, but do not produce biofilm, since the inhibition of gene expression for adhesin als and hwp1, involved in biofilm formation and regulated by *bcr1* [1]. Separation of the filamentation process and biofilm formation showed that the morphogenesis of blastopores to pseudomycelium and consequently the presence of pseudomycelium did not clearly determine the biofilm formation. Pseudomycelium is only a basis, on which under control of the transcription factor bcr1, the adhesins gene gradually express. It is therefore necessary to provide the proper function—filamentous adherence, without which mature biofilm will not be formed. Mutants lacking the activity of tec1, bcr1, als3, or hwp1 proteins exhibit large abnormalities in the biofilm production, which may underline the importance of all these proteins in the early stages of biofilm formation. The ability of residual biofilm formation by these mutants may at the same time indicate that these proteins are not directly involved in adherence to the surface, but in adherence between the blastospore, the mycelium forms, or the adherence mixed between both forms [1, 5, 21].

7. *Candida* communication and the ability to the biofilm formation

For the proper functioning of biofilm, communication between the cells and density regulation is necessary. These tasks are executed by small signaling particles called autoinducers and by responding to the generated signals within population in the *quorum sensing* system. In the culture with a high population density, there are signaling particles which, through diffusion, penetrate other cells running different signals. Exchanging signals lead to specific cell effects and coordination of cellular activity like multicellular behavior. *Candida* yeasts produce several signaling molecules, the accumulation of which determines the development, existence, and breakdown of the biofilm through having a direct influence on the process of mycelial forms creation. The best-known molecule is farnesol ($C_{15}H_{26}O$), a terpene alcohol isolated from *Candida albicans* cells. In the reproduction and maturation phase of the biofilm, the density of cells is relatively small, allowing the formation of mycelium. With the cell concentration increase, the concentration of farnesol, which interacts with the blastospore cell receptors, is increasing, preventing transformation into pseudomycelium and maturation of biofilm. The consequence is the phase of the biofilm dispersion in which individual blastospores and their aggregates are detached [1, 2, 5, 18]. The release of blastospore requires weakening of the adhesive properties, which corresponds to the anti-adhesion ywp1 protein, and the main regulator of the process is the hsp90 protein [5]. Farnesol exogenously inhibits biofilm formation by blocking the expression of many genes responsible for the formation of pseudomycelium and induces expression of the *adh* gene taking part in inhibition of the biofilm formation. The endogenous accumulation of this signaling molecule in biofilm structures may therefore be a factor initiating the breakdown of the biofilm upon reaching a critical cell concentration. Farnesol also influences the expression of ergosterol metabolism genes. At the time of action of azoles blocking the ergosterol synthesis, the substrate to produce this molecule is increased by *Candida albicans*, and the amount of farnesol may even rise 45 times. Farnesol also affects many other processes, such as production of chlamydospores, iron transport, and activation of genes responsible for antibiotic resistance and oxidative stress. Unfortunately, it also has an adverse effect on host cells by inhibiting macrophage activity in the mouse model [1, 2, 5].

Another *Candida albicans* signaling molecule is an autoinducer thiamazole that stimulates the production of mycelium during the intermediate phase of biofilm growth. Comparing to planktonic cells, biofilm cells produce higher amounts of thiamazole [2, 18]. It protects the cells from a decrease in the expression of DNA replication genes, chromosome segregation, and a cell cycle control [1, 2].

The active regulation of the process of detachment from biofilm surface layers, in the state of achieving critical concentration of cells inside, is a crucial role of signaling molecules [1].

8. *Candida* antibiotic resistance

Candida albicans, like most pathogens, developed a number of mechanisms that regulate their virulence. It has developed different strategies to colonize host tissues and break down and weaken its barriers and defense mechanisms. One of the most important virulence factors of *Candida* sp. is the ability to produce mycelial forms that allow a host tissue invasion, at the same time repelling an effective phagocytes attack. The virulence of *Candida* sp. is strongly related to proteins determining of cell integrity, adherence, colonization, or change of phenotypic forms. These proteins are also an effective weapon in the fight against host defense. Most of them are characterized by the presence of anchored glycosylphosphatidylinositol and represent 88% of all covalently bound *Candida albicans* cell membrane proteins. Increasing clinical drug resistance because of abuse of antimicrobial agents is an important phenomenon hindering the fight against these yeasts. *Candida albicans* drug resistance is closely related to the antifungal activity of the drugs used.

Most drug resistance mechanisms to antifungal agents are the results of gene mutations. Usually, these are point mutations of genes encoding drug-binding molecules, enzymes of metabolic pathways, or transcription factors [22]. Such mutations are stable and their acquisition takes time. It is believed that they are the expression of a cell response to chronic stress, for example, resistance-inducing azoles [23] or genetic aneuploidy [24], which changes the expression of multi-drug pump points or transcription factors. Antifungal drugs can also activate a classic, immediate response to a stress. Resistance acquired on this path does not involve the change of genetic material and is reversible, for example, *Candida* sp. phenotype form change or biofilm formation. This reversible change allows us to obtain the time necessary to induce permanent resistance dependent on genetic mechanisms. For the resistance of one of the oldest antifungal agents, 5-FC, the most responsible is uracil phosphoribosyltransferase mutation preventing conversion of 5-fluorouracil to fluorouradine 5-monophosphate [25].

Resistance to polyene, which is still relatively rare today, is obtained by decreasing ergosterol content in the cell membrane, inter alia by *erg3* gene mutation. Lowering ergosterol content in the cell membrane also leads to azole resistance by the increased expression of the erg11p molecule, the azoles binding point. Point mutations of this molecule are responsible for replacing the toxic ergosterol precursors accumulated in the yeast cell by non-toxic ones [25]. *In vitro* studies show different patterns of drug resistance to azoles, frequently overlapping with clinical trials. *In vitro*, the role of the hsp90 molecule chaperone for calcineurin in promoting the rapid acquisition of *Candida albicans* resistance to fluconazole has been identified [26, 27]. Interestingly, the ability to maintain azole resistance even after treatment has ended [28]. Another effective mechanism of azoles resistance is the high expression of multilayer membrane pumps (MFS Mdr1p drug pump or ATP binding cartridge (ABC) of the Cdr1p or Cdr2p pump). These pumps beside azoles are active against a variety of other drugs, apart from echinocandins [25, 29]. Limited resistance to echinocandin is most likely related to their relatively rare use. Although, in recent years, there have been reports both *in vitro* and *in vivo* on *Candida* sp. resistance to echinocandins. The best-known mechanism of resistance to these antibiotic agents is the mutation of the β-1,3-D-glucan synthase Gsc1p subunit [25, 29].

9. *Candida* biofilm and its drug resistance

Particularly dangerous from a clinical point of view is the ability of most clinically important *Candida* species (*Candida albicans, Candida dubliniensis, Candida glabrata, Candida krusei, Candida tropicalis*, and *Candida parapsilosis*) to biofilm formation. The clinical significance of biofilm is increasing with the increasing introduction of various medical devices into the human body. Almost all the contaminations of these devices are the results of their colonization by microorganisms forming organized biofilm structures.

Biofilm *Candida* sp. is characterized by high resistance to all antifungal agents currently used: azoles (fluconazole, itraconazole, voriconazole, posaconazole), echinocandins (kaspofungin, mikafungin, anidulafungin), amphotericin B, flucytosine, but the level of this resistance is different for different drugs. Studies have shown that *Candida* sp. biofilm is resistant to fluconazole at a concentration of 2000 times higher than the MIC value for the planktonic form. The liposomal form of amphotericin B and echinocandin are the most active against *Candida* sp. biofilm. These antibiotic agents exhibit anti-biofilm activity in concentrations 2–25 times higher than the MIC values against planktonic forms [25, 30, 31]. Biofilm resistance is a complex, multi-factor phenomenon that uses the different mechanisms generated by planktonic forms at different stages of biofilm formation. There is also the possibility of generating different mechanisms of drug resistance by individual cell in the biofilm. For example, in the early stages of biofilm-building with low cell concentration, the increase in the activity of drug pumps, lowering the intracellular concentration of azoles, is noted. In mature biofilms characterized by greater cell concentration and many extracellular substances, resistance to amphotericin, azoles, and echinocandins is generated [11, 25, 30]. In addition, it appears that the lower content of ergosterol in mature biofilm is also one of the mechanisms of defense against antifungal agents [22]. The change from planktonic forms to biofilm is a response to unfavorable environmental conditions, which starts a rapid response to stress, which generates, for example, drug resistance. Acquired by acute stress resistance, it is associated with protein kinase activity, calcineurin or hsp90p heat shock protein. Drug-resistant subpopulation protects the pool of cells needed to rebuild the biofilm [25]. An extracellular matrix (ECM) is an important factor in the generation of multi-drug resistance. β-1,3-D-glucan, one of the ECM components, is responsible for drug resistance to fluconazole and amphotericin B. While the role of ECM in generating multi-drug resistance is unquestioned, the mechanisms leading to it remain unexplained.

In the fight against *Candida* biofilm, there are two main problems: a penetration of the drug into the biofilm structure and to overcome the yeast resistance produced by the cells organized in the biofilm. A method of "lock therapy" is conformed to deliver the antifungals directly into the places colonized by the biofilm. To conquer the growing antibiotic biofilm resistance the following strategies are applied: (i) novel antifungal agents in the forms of conjugates, (ii) a multi-drug therapy, (iii) a combination of antifungal agents with nonsteroidal anti-inflammatory drugs, and (iv) agents interfering the communication of cells in the biofilm.

The use of high drug concentrations, for example, higher echinocandin doses used to treat endocarditis, is one of the proposed methods of fighting against *Candida* sp. biofilm. "Lock therapy" uses medical devices (e.g., vascular catheters) for treatment, where high drug doses are introduced into the catheter [32, 33].

Hudson et al. [34] describe a novel form of amphotericin B, dextran aldehyde conjugate with amphotericin B, preservative gel formulation used in local treatment of infections (ligaments, vascular catheters, bones) caused by *Candida* sp. biofilm. *In vitro*, also other compounds: EDTA, ethanol, and high doses of monocycline, are effective in the fight against *Candida* sp. biofilm as "lock therapy" [32, 35].

Another method of fighting infections caused by *Candida* sp. biofilm is the combination therapy of antifungal agents (fluconazole, echinocandin, and amphotericin B) with calcineurin inhibitors such as cyclosporin A or tacrolimus. Such therapy exhibited good *in vivo* activity in the treatment of rat-associated venous catheter infections [35]. Other promising preparations used in "lock therapy" in combination with antifungal agents are compounds that target hsp90 heat shock proteins such as geldanamycin [25, 35]. However, none of these preparations are suitable for systemic use due to their toxicity or lack of confirmed safety in clinical trials.

An interesting proposal seems to be the combination of antifungal preparations with widely used nonsteroidal anti-inflammatory drugs (NSAIDs). Their activity by inhibiting cyclooxygenase prevents yeasts filamentation and thus biofilm formation [32].

Recently, the synergistic effects of 2-adamantanamine, a structural analogue of antiviral amantadine, with fluconazole have been discovered. The mechanism of action is unknown, but it appears that 2-adamantanamine inhibits lanosterol 14-α-demethylase in the ergosterol cycle [32]. The patients' safety of such association has not been established.

Attempts are also being made to use molecules responsible for biofilm communication. One of them is farnesol, which, more than in physiological concentration, leads to biofilm degradation.

Its activity in mouse model studies *in vitro* was comparable to that of azoles. However, *in vivo* studies on animal models have not been conducted [32, 36].

Pulmozyme preparation, comprising recombinant human deoxy ribonuclease (rkDNase), is currently used in inhalation therapy of patients with cystic fibrosis, which targets bacterial biofilm DNA [32].

10. Probable environmental circulation of *Candida* strains

Besides the most frequent fungal pathogen *Candida albicans*, non-*albicans Candida* strains are isolated from the patients and clinical environments. Among non-*albicans*, the common clinical isolates are *Candida glabrata, Candida lusitaniae, Candida parapsilosis, Candida auris, Candida tropicalis*, and *Candida dubliniensis*. *Candida* sp. are the widespread yeasts in food products serving as natural flora members or food-contaminants. The examples of food-associated *Candida* yeasts are *Candida lusitaniae, Candida famata, Candida parapsilosis, Candida colliculosa, Candida tropicalis, Candida krusei, Candida boidinii*, and *Candida pelliculosa*. Considering the possibilities of *Candida* strains' natural circulation between food and clinical environments, a question arises if the food-borne strains can be a threat for specific groups of patients. Our previous work presented the examples of antibiotic resistance of a variety of *Candida* clinical and food-borne isolates [37]. Within the study, 24 clinical strains of *Candida albicans* and 1 *Candida glabrata* strain as well as a *Candida lusitaniae* strain were compared with 18 non-*albicans* food-borne candidas. The set of food-borne isolates consisted of *Candida lusitaniae* (four strains), *Candida famata* (two strains), *Candida parapsilosis* (one strain), *Candida colliculosa* (one strain), *Candida tropicalis* (one strain), *Candida krusei* (four strains), *Candida boidinii* (three strains), *Candida rugosa* (one strain), and *Candida pelliculosa* (one strain). The strains sensitivity to the nystatin (polyenes), fluconazole (triazoles I generation), voriconazole (triazoles II generation), and caspofungin (echinocandins) were checked. It was found that all the tested strains were sensitive to caspofungin but 15 strains differed in sensitivity to nystatin, fluconazole, and voriconazole irrespective of their origin. Interestingly, two of four tested food-borne strains of *Candida krusei* were not susceptible to fluconazole, and the third one was classified as intermediate. All *Candida krusei* isolates were sensitive to fluconazole. One clinical isolate of *Candida glabrata* was not sensitive to fluconazole. Triazoles were the last effective not totally inhibiting the growth of the clinical isolates and five food-borne strains. The results proved that a group of strains expressing intermediate sensitivity or resistance to the tested antibiotics include both clinical and food-borne isolates.

According to the biochemical profiles, the tested strains were classified in two groups: (i) 24 *Candida albicans* clinical isolates and 1 strain of food-borne yeast *Candida tropicalis*, which was isolated from pickled cucumbers; (ii) 17 food-borne strains and 2 clinical isolates *Candida glabrata* and *Candida lusitaniae*. What is more, *Candida albicans* isolates expressed vast biochemical heterogeneity. A yeast adaptation to the host organism may explain these differences.

Both *Candida albicans* and *Candida glabrata*, typical human pathogens, were not found in food [37]. The noted biochemical profiles conformity of some clinical and food-borne isolates may be a result of *Candida* transmission via food. The similarity of food-borne *Candida tropicalis* to the *Candida albicans* strains isolated from clinical patients implies the possibility of circulating of antibiotic-resistant strains outside the hospital environment and the possible yeast infection caused by yeasts entered into the body with food.

The plasticity of *Candida* yeasts subjected to non-conventional antifungal compounds like essential oils were also proved [38, 39]. Both *Candida albicans* and food-borne isolates, *Candida rugosa, Candida famata,* and *Candida krusei,* have changed their properties at the presence of tea tree oil (*Melaleuca alternifolia* Maiden & Betche Cheel), thyme oil (*Thymus vulgaris* L.), and clove oil (*Syzygium aromaticum* L. Merr. & L.M. Perry).

11. Conclusions

Biofilm-forming microorganisms, including *Candida* species, are characterized by high invasiveness, the ability to cause dangerous, and difficult to treat infections. Furthermore, the cells in the biofilm, compared to planktonic forms, show reduced sensitivity to chemical compounds with antifungal activity and increased survival under unfavorable environmental conditions. The morphological diversity of the biofilm structures formed by *Candida albicans* and non-*albicans* strains allows these yeasts to colonize both biotic and abiotic surfaces. The emergence of antifungal resistance with the development of biofilms is still a problem. The incidences of medical equipment colonization by *Candida* yeasts are constantly noted. Moreover, the proven biochemical profiles conformity of some clinical and food-borne isolates may be a result of *Candida* transmission via food.

Author details

Marta Targalska and Alina Kunicka-Styczyńska*

*Address all correspondence to: alina.kunicka@p.lodz.pl

Institute of Fermentation Technology and Microbiology, Lodz University of Technology, Lodz, Poland

References

[1] Mnichowska-Polanowska M, Kaczała M, Gierdys-Kalemba S. Charakterystyka biofilmu *Candida*. Mikologia Lekarska. 2009;**16**:159-164

[2] Kołwzan B. Analiza zjawiska biofilmu—Warunki jego powstawania i funkcjonowania. Ochrona Środowiska. 2011;**33**:3-14

[3] Budzyńska A, Różalska B. Potencjalne wykorzystanie roślinnych olejków eterycznych w zwalczaniu zakażeń z udziałem biofilmów drobnoustrojów. Życie Weterynaryjne. 2012;**87**:213-215

[4] Matejczyk M, Suchowierska M. Charakterystyka zjawiska *quorum sensing* i jego znaczenie w aspekcie formowania i funkcjonowania biofilmu w inżynierii środowiska, budownictwie, medycynie oraz gospodarstwie domowym. Budownictwo i Inżynieria Środowiska. 2011;**2**:71-75

[5] Nawrot U. Znaczenie biofilmu w patogenezie i leczeniu grzybic. Zakażenia. 2013;**4**:56-59

[6] Mazaheritehrani E, Sala A, Orsi CF, Neglia RG, Morace G, Blasi E, et al. Human pathogenic viruses are retained in and released by *Candida albicans* biofilm *in vitro*. Virus Research. 2014;**179**:153-160

[7] Shao X, Cao B, Xu F, Xie S, Yu D, Wang H. Effect of postharvest application of chitosan combined with clove oil against citrus green mold. Postharvest Biology and Technology. 2015;**99**:37-43

[8] Reśliński A, Mikucka A, Szczęsny W, Szmytkowski J, Gospodarek E, Dąbrowiecki S. Wykrywanie biofilmu *in vivo* na powierzchni siatki chirurgicznej—Opis przypadku. Chirurgia Polska. 2008;**10**:181-188

[9] Sadowska B, Budzyńska A, Więckowska-Szakiel M, Paszkiewicz M, Stochmal A, Moniuszko-Szajwaj B, et al. New pharmacological properties of *Medicago sativa* and *Saponaria officinalis* saponin-rich fractions addressed to *Candida albicans*. Journal of Medical Microbiology. 2014;**63**:1076-1086

[10] Rosseti IB, Rochab JBT, Costa MS. Diphenyl diselenide (PhSe)2 inhibits biofilm formation by *Candida albicans*, increasing both ROS production and membrane permeability. Journal of Trace Elements in Medicine and Biology. 2015;**29**:289-295

[11] Dorocka-Bobkowska B, Konopka K. Biofilm formation by *Candida* and its role in the pathogenesis of chronic infections—Review. Dental and Medical Problems. 2003;**40**:405-410

[12] Czaczyk K. Czynniki warunkujące adhezję drobnoustrojów do powierzchni abiotycznych. Postępy Mikrobiologii. 2004;**43**:267-283

[13] Karatan E, Watnick P. Signals, regulatory networks, and materials that build and break bacterial biofilm. Microbiology and Molecular Biology Reviews. 2009;**73**:310-347

[14] Strużycka I, Stępień I. Biofilm—Nowy sposób rozumienia mikrobiologii. Nowa Stomatologia. 2009;**3**:85-89

[15] Gospodarek E. *Quorum sensing*—Chemiczne komunikowanie się drobnoustrojów. Postępy Mikrobiologii. 2004;**43**(S):12

[16] Sanchez-Vargas LO, Estrada-Barraza D, Pozos-Guillen AJ, Rivas-Caceres R. Biofilm formation by oral clinical isolates of *Candida* species. Archives of Oral Biology. 2013; **58**:1318-1326

[17] Rossi BP, Garcia C, Alcaraz E, Franco M. *Stenotrophomonas maltophilia* interferes via the DSF-mediated *quorum sensing* system with *Candida albicans* filamentation and its planktonic and biofilm models of growth. Revista Argentina de Microbiología. 2014;**46**:288-297

[18] Tsai PW, Chen YT, Yang CY, Chen HF, Tan TS, Lin TW, et al. The role of Mss11 in *Candida albicans* biofilm formation. Molecular Genetics and Genomics. 2014;**289**:807-819

[19] Samaranayake YH, Cheung BPK, Yau JYY, Yeung SKW, Samaranayake LP. Human serum promotes *Candida albicans* biofilm growth and virulence gene expression on silicone biomaterial. PLoS One. 2013;**8**:e62902

[20] Orsi CF, Borghi E, Colombari E, Neglia RG, Quaglino D, Ardizzoni A, et al. Impact of *Candida albicans* hyphal wall protein 1 (HWP1) genotype on biofilm production and fungal susceptibility to microbial cells. Microbial Pathogenesis. 2014;**69-70**:20-27

[21] Connolly LA, Riccombeni A, Grozer Z, Holland LM, Lynch DB, Andes DR, et al. The APSES transcription factor Efg1 is a global regulator that controls morphogenesis and biofilm formation in *Candida parapsilosis*. Molecular Microbiology. 2013;**90**:36-53

[22] Ferreira C, Silva S, Oliveira FF, Pinho E, Henriques M, Lucas C. *Candida albicans* virulence and drug-resistance requires the O-acyltransferase Gup1p. Microbiology. 2010;**10**:238-251

[23] Perepnikhatka V, Fischer FJ, Niimi M, Baker RA, Cannon RD, Wang YK, et al. Specific chromosome alterations in fluconazole-resistant mutants of *Candida albicans*. Journal of Bacteriology. 1999;**181**:4041-4049

[24] Selmecki A, Forche A, Berman J. Aneuploidy and isochromosome formation in drug-resistant *Candida albicans*. Science. 2006;**313**:367-370

[25] Cannon RD, Lamping E, Holmes AR, Niimi K, Tanabe K, Niimi M, et al. *Candida albicans* drug resistance—Another way to cope with stress. Microbiology. 2007;**153**:3211-3217

[26] Cowen LE, Carpenter AE, Matangkasombut O, Fink GR, Linquist S. Genetic architecture of Hsp90-dependent drug resistance. Eukaryotic Cell. 2006;**5**:2184-2188

[27] Cowen LE, Lindquist S. Hsp90 potentiates the rapid evolution of new traits: Drug resistance in diverse fungi. Science. 2005;**309**:2185-2189

[28] Anderson JB. Evolution of antifungal-drug resistance: Mechanisms and pathogen fitness. Nature Reviews. Microbiology. 2005;**3**:547-556

[29] Baixench MT, Desnos-Ollivier M, Garcia-Hermoso D, Bretagne S, Ramires S, Piketty C, et al. Acquired resistance to echinocandins in *Candida albicans*: Case report and review. The Journal of Antimicrobial Chemotherapy. 2007;**59**:1076-1083

[30] Kojic EM, Darouiche RO. *Candida* infections of medical devices. Clinical Microbiology Reviews. 2004;**17**:255-267

[31] Heizmann P, Klefisch F, Heizmann WR. Basic research—Significance of detection and clinical impact of *Candida albicans* in non-immunosupressed patients. Pharmacology & Pharmacy. 2011;**2**:354-360

[32] Walvaren CJ, Leea SA. Antifungal lock therapy. Antimicrobial Agents and Chemotherapy. 2013;**57**:1-8. DOI: 10.1128/AAC.01351-12

[33] DiMondi VP, Townsend ML, Johnson M, Durkin M. Antifungal catheter lock therapy for the management of a persistent *Candida albicans* bloodstream infection in an adult receiving hemodialysis. Pharmacotherapy. 2014;**34**:120-127

[34] Hudson SP, Langer R, Fink GR, Kohane DS. Injectable *in situ* cross-linking hydrogels for local antifungal therapy. Biomaterials. 2010;**31**:1444-1452

[35] Nett JE. Future directions for anti-biofilm therapeutics targeting *Candida*. Expert Review of Anti-Infective Therapy. 2014;**12**:375-382

[36] Ramage G, Saville SP, Wickes BL, Lopez-Ribot JL. Inhibition of *Candida albicans* biofilm formation by farnesol, a *quorum sensing* molecule. Applied and Environmental Microbiology. 2002;**68**:5459-5463

[37] Maroszyńska M, Kunicka-Styczyńska A, Rajkowska K, Maroszyńska I. Antibiotic sensitivity of *Candida* clinical and food-borne isolates. Acta Biochimica Polonica. 2013;**60**:719-724

[38] Rajkowska K, Kunicka-Styczyńska A, Maroszyńska M, Dąbrowska M. The effects of thyme and tea tree oils on morphology and metabolism *Candida albicans*. Acta Biochimica Polonica. 2014;**61**:305-310

[39] Rajkowska K, Kunicka-Styczyńska A, Pęczek M. Hydrophobic properties of *Candida* spp. under the influence of selected essential oils. Acta Biochimica Polonica. 2015;**62**:663-668

Chapter 3

Multidrug Sensitive Yeast Strains, Useful Tools for Chemical Genetics

Takumi Chinen, Keisuke Hamada, Akihiro Taguchi, Yukihiro Asami, Kazuro Shiomi, Yoshio Hayashi and Takeo Usui

Additional information is available at the end of the chapter

http://dx.doi.org/10.5772/intechopen.70664

Abstract

The budding yeast *Saccharomyces cerevisiae* is a useful eukaryote model organism for application to chemical biology studies, for example, drug screening, drug evaluation, and target identification. To use yeast for chemical biology research, however, it has been necessary to construct yeast strains suitable for various compounds because of their high drug resistance. Hence, the deletion of all multidrug resistance genes except for those that are important for viability and for genetic experiments/manipulation could increase the drug sensitivity without influencing the transformation, mating, or sporulation efficiency. There are two major factors conferring multidrug resistance in *S. cerevisiae*: one is the drug efflux system and the other is the permeability barrier. We therefore constructed a strain which shows high sensitivity to multiple drugs by disrupting the drug efflux system using ATP-binding cassette transporters and suppressing the membrane barrier system by introducing an ERG6-inducible system. In this review, we discuss the construction of our multidrug-sensitive yeast strains and their application in chemical biology.

Keywords: multidrug-sensitive yeast, drug efflux system, permeability barrier system, drug target identification, drug screening

1. Introduction

1.1. Screening and target identification of bioactive small molecules: important processes in chemical genetics

The screening of bioactive small molecule compounds is the most important process in drug development. Natural products which have structural diversity isolated from microorganisms, plants, and animals are useful sources in the field of drug development [1]. Structurally, new natural products might show novel activities such as antimicrobial, antiviral, and antitumor activities. These natural products also provide useful information for medicinal chemistry, and allow the development of new synthetic compounds as novel medicines. For example, eribulin, a semi-synthetic derivative of halichondrin B, has been approved as an anti-cancer drug [2–4]. Therefore, the screening and identification of new small molecules open new avenues for drug development. There are two major ways to identify bioactive small molecules: phenotypic screening and target-based screening. Phenotypic screening is based on cytotoxicity [5–7], cell cycle arrest [8], immune-suppression [9], and morphological changes [10] of drug-treated cells, fungi, and bacteria. Target-based screening is performed based on measurable readouts such as enzymatic activity inhibition [11] or drug-protein interaction [12]. These approaches have identified useful small molecules and medicines.

Target identification (Target ID) of small molecules is also quite important in order to develop safe and useful drugs [13]. Thalidomide, a cautionary example, was used as a sedative a half-century ago before it was found to be teratogenic and to cause multiple birth defects [14]. However, thalidomide is also used in the treatment of Hansen's disease, myeloma [14], and so on. In addition, immunomodulatory drugs derived from thalidomide have been developed as a new class of anti-cancer drugs and novel medicines for treating ribosomopathies such as 5q-syndrome [15]. Recently, cereblon, a substrate receptor of the CRL4 E3 ubiquitin ligase, has been identified as a primary target of thalidomide teratogenic [16] and anti-cancer [15] activity. These lines of research provide useful information that cereblon may pose a risk of teratogenic activity and simultaneously serve as an attractive molecular target for immunomodulatory drug development. To identify the relevant target molecules and target pathways, indirect and direct approaches have been used [13]. The indirect approaches include phenotypic analysis and large-scale analysis such as proteomic and genome-wide analyses. Some specific changes in cell morphology, cell cycle arrest, and other phenotypes provide us useful information for predicting targets of the drugs. Based on this property, Morphobase, an encyclopedic database of the morphological changes that occur in drug-treated cells, has been constructed and applied to drug target discovery [17]. Large-scale analyses such as proteomics, metabolomics, and transcriptome analysis of drug-treated cells have been performed to predict the target pathways of bioactive small molecules [18]. Genome-wide genetic studies are also frequently used for drug target ID. For example, synthetic lethal/sick genetic interaction analyses [19, 20], genome-wide overexpression screening [21], and haploinsufficiency-chemical sensitive assays [22] have been used to analyze the mode of action of various drugs. On the other hand, direct approaches, such as affinity probe approaches and genetic analyses, are quite useful to identify the direct target molecules of drugs. By using affinity probe approaches, the targets of thalidomide [16] and FK506 [23] have been identified. Genetic analysis is another powerful method of identifying not only drug targets [24–29] but also the signaling pathway affected by a drug. Genetic studies using model organisms such as yeast have contributed to identification of the target molecules of bioactive compounds.

The identification of new bioactive small molecules and elucidation of their target molecules/signaling pathways are important not only for developing medicines but also for basic science. Such compounds are a useful tool for understanding the fundamental protein

http://dx.doi.org/10.5772/intechopen.70664

functions in cells. Well-known examples are famous immunosuppressants such as FK506, cyclosporine, and rapamycin. These compounds inhibit immunophilin and T-cell activation through different mechanisms [30]. Studies of these compounds have revealed their detailed immunoreaction mechanisms [30]. Mitotic inhibitors are another example. Mitotic spindle formation and chromosome segregation are fast processes that are completed within approximately 1 hour. Therefore, by taking advantage of rapid pharmacological intervention, studies using microtubule inhibitors (αβ-tubulin inhibitors [31–33] or γ-tubulin inhibitor [12]), mitotic kinesins (Eg5 [34, 35]), and mitotic kinase inhibitors (aurora kinases [36, 37], Cdk1 [38], Plk1 [39, 40], Mps1 [41, 42]) highlighted useful information regarding the temporal regulation of mitotic spindle architecture and faithful chromosome segregation. These findings could in turn contribute to further drug development. Therefore, target ID of newly found useful bioactive compounds is quite an important process in both basic science and medicine development.

1.2. *Saccharomyces cerevisiae*, a useful model organism for chemical genetics

Saccharomyces cerevisiae is one of the most frequently used model organisms in chemical genetics. The properties of *S. cerevisiae* along with easy-to-use genetic analyses, mutational analyses, gene disruption, and genome modification have facilitated both chemical screening and target ID (**Table 1**). For example, the target of rapamycin (TOR) has been found by genetics using *S. cerevisiae* [29]. In addition, *S. cerevisiae* is useful for chemical screening [43, 44]. However, *S. cerevisiae* generally shows higher resistance against various compounds compared with mammalian cells, except in the case of a few compounds such as rapamycin (**Table 2**). This disadvantage limits the application of *S. cerevisiae* in chemical screening. Therefore, *S. cerevisiae* showing sensitivities against drug of interest has been quite useful. For example, *S. cerevisiae* quadruple deletion mutant lacking *yrr1*, *yrs1*, *pdr1*, and *pdr3* was constructed for the analyses of target molecule of reveromycin A. However, construction of sensitive yeast suitable for each compound is a time-consuming process. To overcome this drawback, we developed two multidrug-sensitive strains which have proven quite useful for

Compound	Approach	Finding	Ref.
Benomyl	Pathway analysis	Identification of Mad1, Mad2, Mad3 as mitotic spindle checkpoint proteins by using benomyl sensitive mutants	[31]
Benomyl	Pathway analysis	Identification of Bub1, Bub2, Bub3 as mitotic spindle checkpoint proteins by using benomyl sensitive mutants	[32]
Reveromycin A	Target ID	Identification of *ILS1* as a target of reveromycin A	[27]
Curvularol	Target ID	Identification of *RPL3* as a target of curvularol	[28]
Rapamycin	Target ID	Identification of *TOR* as a target of rapamycin	[29]
Eudistomin C	Target ID	Identification of *RPS14* as a target of eudistomin C	[50]
Splitomicin	Screening	Identification of splitomicin as a NAD^+-dependent histone deacetylase inhibitor	[51]

Table 1. The examples of chemical genetics studies using ***S. cerevisiae***.

	Mammalian cell line (HeLa)	Budding yeast (BY4741)
Cycloheximide (μM)	0.2	270
Digitonin (μM)	0.4	1.9
Fluphenazine (μM)	13	51
Latrunculin A (nM)	0.2	>240
4-Nitroquinoline 1-oxide (μM)	0.1	7.1
Rapamycin (nM)	>300	7.1
Staurosporine (μM)	0.1	15.1
Tunicamycin (μM)	1.8	>120

HeLa cells (3×10^3 cells/well in 96 well plate) and BY4741 cells (3.8×10^5 cells/well in 96 well plate) were treated with various concentrations of compounds for 48 and 8 h, respectively. Cell viabilities were determined by WST-8 (Dojindo, Kumamoto, Japan) and IC50 values were calculated.

Table 2. The IC50 values of compounds against HeLa cells and *S. cerevisiae*.

research in chemical biology. There are two major systems conferring multidrug resistance in *S. cerevisiae*: one is the drug efflux system, which exports drugs into vacuoles or outside of cells, and the other is the permeability barrier, which blocks the penetration of drugs into the cells (**Figure 1**). The drug efflux system consists of ATP-binding cassette (ABC) transporters that export xenotoxic compounds outside of cells or inside of vacuoles, and their transcriptional factors [45–47]. *S. cerevisiae* has at least 16 ABC transporters, of which Pdr5p, Snq2p, and Yor1p confer multidrug resistance by exporting bioactive small molecules out of cells. Four transcriptional factors (Pdr1p, Pdr3p, Pdr8p, and Yrr1p) up-regulate the transcription of most of the ABC transporters [45–47]. A permeability barrier is conferred by ergosterol in the yeast plasma membrane. Therefore, ABC transporter-related genes and ergosterol synthesis genes were frequently disrupted to construct drug-sensitive strains. For instance, a strain in which *pdr1*, *pdr3* (genes encoding transcriptional factors for ABC transporters), and *erg6* (a gene involved in ergosterol synthesis) were disrupted was used for drug screening [43]. However, the *erg6* deletion mutant shows decreased transformation and sporulation efficiencies that are essential for yeast genetic analysis. In addition, some of the transporters located in the vacuole membrane are involved in the detoxination of metabolites as well as xenotoxins, and their disruption results in growth defects. Therefore, to make a yeast strain sensitive to a wide range of drugs, it is necessary to suppress both efflux and barrier systems without affecting the genetic properties and growth rate. Hence, we speculated that the disruption of all ABC transporters located on the plasma membrane that are not important for viability and genetic experiments or for the conditional expression regulation of the *ERG6* gene could increase the drug sensitivity without influencing the transformation, mating, or sporulation efficiency.

In this review, we discuss the construction of two multidrug-sensitive yeast strains, 12geneΔHSR [48] and 12geneΔHSR-iERG [49], which are available for genetic analysis. We also discuss the application of these strains in drug screening and target ID [50].

http://dx.doi.org/10.5772/intechopen.70664

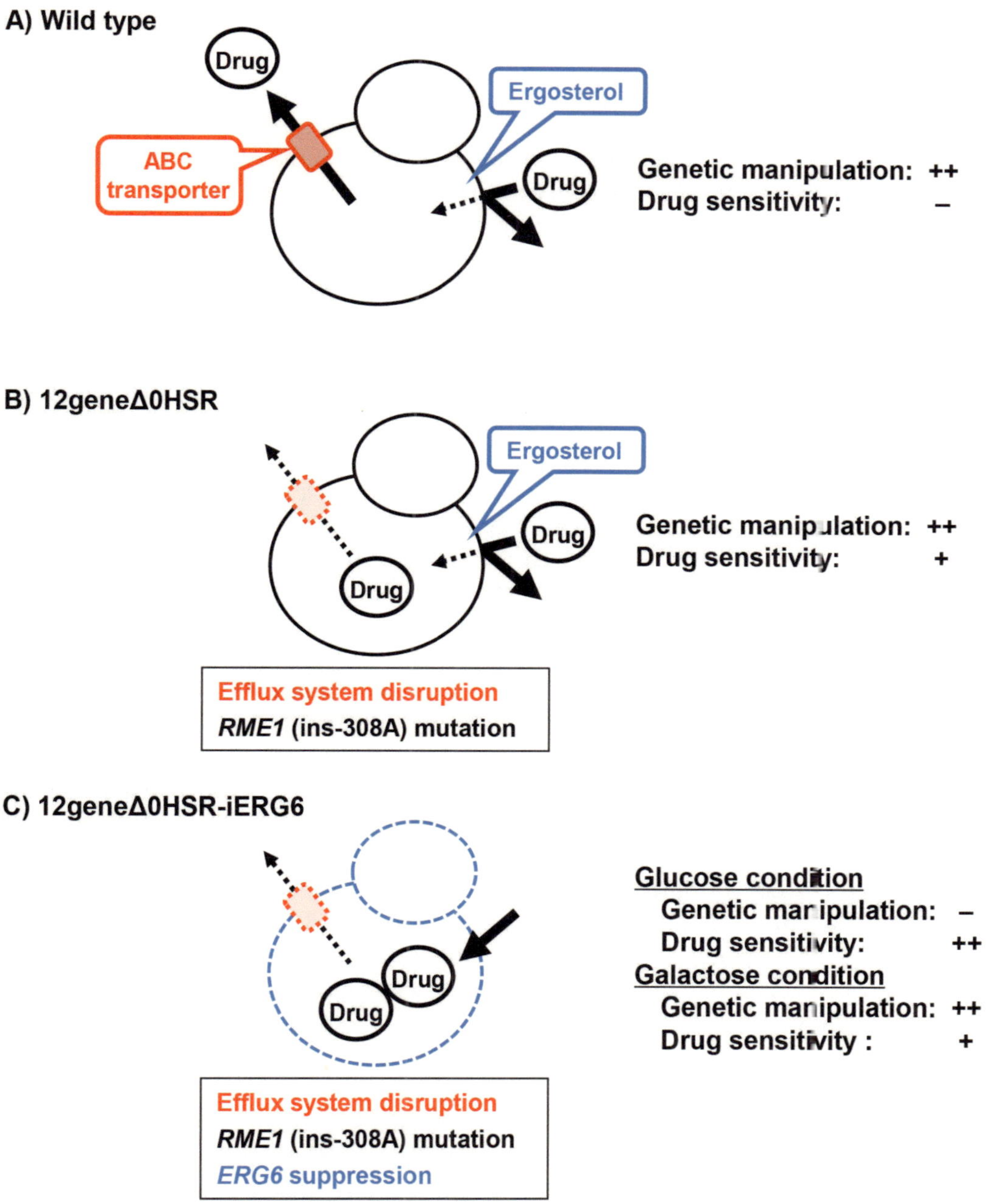

Figure 1. The work flow of the construction of multidrug-sensitive strains. (A) The parental strain, BY4741, possesses high genetic manipulation availability, but shows high drug resistance. (B) 12geneΔ0HSR, created by disruption of the drug efflux system and introduction of the *RME1(ins-308A)* mutation, achieves drug-sensitivity without compromising the genetic manipulation availability. (C) 12geneΔ0HSR-iERG was created by the insertion of a *gal1* promoter into *ERG6*. This strain shows high drug sensitivity but drastically decreased genetic manipulation availability under the glucose condition, because ERG6p expression is repressed. Instead, genetic manipulation is available under the galactose condition through enhancement of the ERG6p expression.

2. Construction and application of multidrug-sensitive yeast strains

2.1. Construction of multidrug-sensitive yeast strains

We constructed a multidrug-sensitive yeast strain by disrupting 12 ABC transporter-related genes and suppressing the *ERG6* gene. The work flow is shown in **Figure 1**. As a first step, we focused on drug efflux systems. The drug efflux system composed of ABC transporters confers resistance against a wide variety of compounds [45–47]. Therefore, it is difficult to predict which transporters will confer drug resistance against the drug of interest. We thus decided to construct the 12geneΔ0 strain through the disruption of all of the ABC transporters involved in drug export located on the plasma membrane and transcription factors involved in multidrug resistance specifically on a BY4741 background [48]. Gene disruption of eight gene-encoding ABC transporters (*AUS1, PDR5, PDR10, PDR11, PDR12, PDR15, SNQ2,* and *YOR1*) and four genes encoding transcriptional factors (*PDR1, PDR3, PDR8,* and *YRR1*) was carried out using a PCR-based markerless gene disruption method modified from the *delitto perfetto* method [52]. Because 12geneΔ0 leaves no marker genes in the genome, auxotroph markers which the parental strain originally possesses can be used for further studies. To use 12geneΔ0 for chemical genetics, it is important to show not only its multidrug sensitivity but also its transformation, mating, and sporulation efficiencies, which are necessary for genetic analysis. The transformation and mating efficiency of 12geneΔ0 were on the same order as those of the parental strain BY4741 (**Table 3**). However, the sporulation efficiency was drastically decreased in 12geneΔ0 (**Table 3**). It was reported that single-nucleotide polymorphisms of three genes (a noncoding regulatory region of *RME1(ins-308A)*, and two missense mutations in *TAO3* and *MKT1*) are involved in sporulation efficiency, and when these mutations were introduced in S288c, the parental strain of BY4741, the sporulation efficiency increased [53]. We therefore introduced the *RME1(ins-308A)* and *MKT1(D30G)* mutations into 12geneΔ0. Although both mutations increased the sporulation efficiencies, the *MKT1(D30G)* mutant formed petite colonies as reported previously [54]. Therefore, we decided to use the *RME1* mutant for our studies, and the strain created was named 12geneΔ0HSR (12geneΔ0 strain showing High Sporulation by *RME1(ins-308A)* mutation) [48]. 12geneΔ0HSR showed sporulation efficiency comparable to that of BY4741. By testing the drug sensitivities of the 12geneΔ0HSR, BY4741*Δerg3*, and BY4741*Δerg6* strains, we revealed that there are different spectrums of drug resistance conferred by the efflux and barrier systems (**Figure 2**) [48], suggesting that it is necessary to disrupt both the drug efflux and permeability barrier systems to make a strain with high sensitivity against a wide range of multiple drugs. To disrupt the permeability barrier system without affecting any of the genetic properties, we introduced the conditional expression promoter *GAL1p* in the *ERG6* gene in 12geneΔ0HSR (**Figure 1**) [49]. The constructed strain, 12geneΔ0HSR-iERG6, showed improved sensitivities to several compounds under the glucose condition (*ERG6* suppression), and it exhibited sufficient transformation and sporulation efficiencies under the galactose condition (*ERG6* expression) (**Table 3**). Because of its high sensitivities to several compounds, the 12geneΔ0HSR-iERG6 strain will be a useful tool in chemical biology studies.

http://dx.doi.org/10.5772/intechopen.70664

	Transformation efficiency (Cfu/μg)	Mating efficiency (%)	Sporulation efficiency (%)
BY4741	$9.6 \times 10^5 \pm 2.2 \times 10^5$	17.7 ± 7.5	21.9 ± 6.8
Δerg6	55.0 ± 51.3	4.8 ± 1.7	9.4 ± 4.7
12geneΔ0	$1.2 \times 10^5 \pm 2.0 \times 10^4$	15.7 ± 5.3	5.0 ± 2.9
12geneΔ0HSR	N.D.	N.D.	28.8 ± 4.6
12geneΔ0HSR-iERG6 (under glucose condition)	7.0 ± 8.2	6.4 ± 2.2	0.0 ± 0.0
12geneΔ0HSR-iERG6 (under galactose condition)	$3.0 \times 10^4 \pm 2.4 \times 10^4$	N.D.	10.7 ± 3.0

Values are mean ± S.D. calculated from three independent experiments. These data are edited from **Figure 1** of Ref. [48] for BY4741, *Δerg6*, 12geneΔ0, and 12geneΔ0HSR, or **Figure 2** of Ref. [49] for 12geneΔ0HSR-iERG6.

Table 3. Comparison of the efficiencies of transformation, mating and sporulation between BY4741, ***erg6*** disruptant and 12geneΔ0HSR.

2.2. Application 1: drug screening

2.2.1. Availability of 12geneΔ0HSR-iERG6 in drug screening

In general, *S. cerevisiae* exhibits high levels of drug resistance, which is an obstacle for drug screening. In fact, most of the compounds used for clinical or basic research show higher IC_{50} values against *S. cerevisiae* than against mammalian cells (**Table 2**). Therefore, multidrug-sensitive strains of *S. cerevisiae*—for example, the *pdr1 pdr3 erg6* triple mutant or *pdr1 pdr3 yrs1 yrr1* quadruplex mutant—have been used for drug screening [43, 55]. To test the superiority of our strain, we screened mitochondrial inhibitors from microbial secondary metabolites and compared the hit ratio of 12geneΔ0HSR-iERG6 with that of BY25929 (*yrs1::HIS3 yrr1::TRP1 pdr1::hisG pdr3::hisG*), a multidrug-sensitive quadruplex mutant (**Tables 4** and **5**).

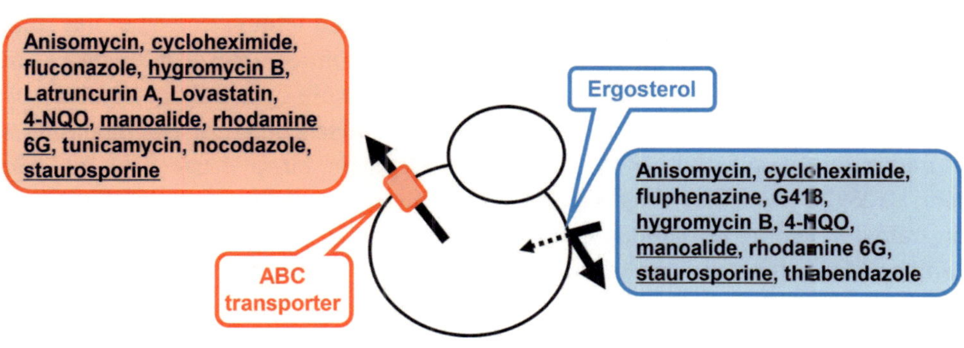

Figure 2. Drugs to which resistance was conferred by ABC transporters, ergosterol or both systems (indicated by underlining), respectively.

	Number of broth	Number of hit broth	Hit ratio (%)
Origin			
Fungus	2664	149	5.6
Actinomycetes	5617	289	5.1
Total	8281	438	5.3

Table 4. Hit ratio of screening of mitochondrial inhibitor using quadruplex mutant, BY25929.

	Number of broth	Number of hit broth	Hit ratio (%)
Origin			
Fungus	3144	270	8.6
Actinomycetes	3067	253	8.2
Total	6211	523	8.4

Table 5. Hit ratio of screening of mitochondrial inhibitor using 12geneΔ0HSR-iERG6.

To identify the mitochondrial inhibitors, we used the difference in cell growth between the glucose medium and the glycerol medium. Yeast can use glycerol as a respiratory substance after the conversion to dihydroxyacetone phosphate via glycerol-3-phosphate by cytosolic and mitochondrial enzymes, GUT1p and GUT2p, respectively. Therefore, yeast could grow even in the presence of a mitochondrial inhibitor in glucose medium because of anaerobic respiration, but not in glycerol medium in which one of the metabolites in glycolysis, dihydroxyacetone phosphate, could not be produced. Therefore, we compared the growth inhibition induced by microbial broth samples on glucose medium (1% yeast extract, 2% polypeptone, 2% glucose, 1.5% agar) with that on glycerol medium (1% yeast extract, 2% polypeptone, 3% glycerol, 1.5% agar), and chose the broth which inhibited yeast growth on glycerol medium but not on glucose medium [55]. Growth inhibition activities of microbial broth samples were evaluated using the paper disc method on agar plates inoculated with recombinant *S. cerevisiae* strains. In detail, 6 mm sterile filter discs impregnated with each compound solution (10 μl) were placed on the agar plate using a forceps (medium volume; 30 ml/plate, cell number; 1.5×10^6 cells/plate, plate dimension; $144 \times 100 \times 14.5$ mm, square shape), and the plates were incubated at 30°C for 48 h. After incubation, the diameters of the zone of inhibition were measured with a vernier caliper. As shown in **Table 4**, the hit ratio using the quadruplex mutant, BY25929, was about 5%. Because the hit ratio when wild-type yeasts (W303-derived yeast strains) were used in a similar screening system was 1.4% (fungus samples 0.5% (44 total hits among 8610 samples), actinomycetes samples 3.2% (125 total hits among 3912 samples), this result suggests that the quadruplex mutant is useful for drug screening with a high hit ratio. Indeed, a novel compound, decatamariic acid, was isolated as a mitochondrial inhibitor using the quadruplex mutant [55]. Moreover, the hit ratio using 12geneΔ0HSR-iERG6 increased to about 8% (**Table 5**).

To determine whether it is possible to isolate the novel compounds or not, we selected the microbial broths which were detected using 12geneΔ0HSR-iERG6 but not using the quadruplex mutant. We found a total of 46 broths (fungus origin: 16 broths; actinomycetes origin: 30 broths) which inhibited the growth of 12geneΔ0HSR-iERG6 specifically. Among these broths, we selected two fungus broths for further purification of active metabolites, and isolated 4,6′-anhydrooxysporidinone (**1**, fusoxypyridone [56]), pestalotic acid A (**2**), and three novel compounds (manuscript in preparation) (**Figure 3**). 4,6′-Anhydrooxysporidinone has been isolated from *Fusarium oxysporum* in the course of the screening of anti-angiogenesis inhibitors [57], but showed weak cytotoxicity against mammalian cell cultures (IC_{50} > 100 μM) and anti-MRSA activity (MIC = 100 μg/ml) [58]. Pestalotic acid A has been isolated from a *Pestalotiopsis* sp. as an antimicrobial compound containing a furylidine tetronic acid core [59]. Because of the lack of biological activity other than antimicrobial activities, the observation of antifungal activity is a novel insight. These results strongly suggest that 12geneΔ0HSR-iERG6 would be useful for drug screening.

2.2.2. *Screening of readthrough compounds*

Because the usefulness of our strains was confirmed, we next performed the preliminary screening of compounds that show readthrough activities. Readthrough compounds allow the translational machinery to skip nonsense mutations encoding premature termination codons (PTCs) and could become medicines for hereditary diseases caused by PTCs (**Figure 4**). To date, many small molecules have been developed as readthrough drug candidates. Several forms of aminoglycoside antibiotics, such as gentamicin (**3**), G418 (**4**), and its analogues, have been reported to show readthrough activities (**Figure 5**) [60]. Barton-Davis *et al.* revealed that the dystrophin expression in *mdx* mice, an animal model of duchenne muscular dystrophy (DMD) is increased after the administration of gentamicin (**3**) [61]. Novel aminoglycosides derived from gentamicin, which showed readthrough activity against four different nonsense DNA constructs underlying genetic diseases, were also recently reported [62]. However, long-term treatment with aminoglycosides showed serious side effects such as nephrotoxicity [63] and ototoxicity [64]. As a non-aminoglycoside readthrough compound, ataluren (**5**), which is a 1,2,4-oxadiazole derivative developed from a chemical library, promotes dystrophin production in primary muscle cells from humans and *mdx* mice (**Figure 5**) [65]. It was also found that (+)-negamycin (**6**), which is a dipeptide-like antibiotic containing a hydrazide

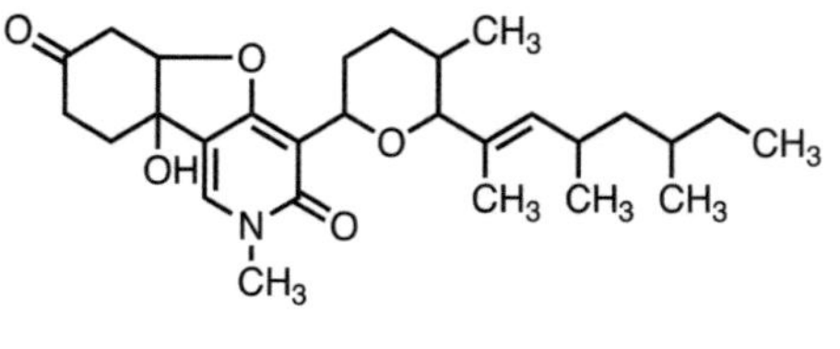

4,6'-Anhydrooxysporidinone (**1**)

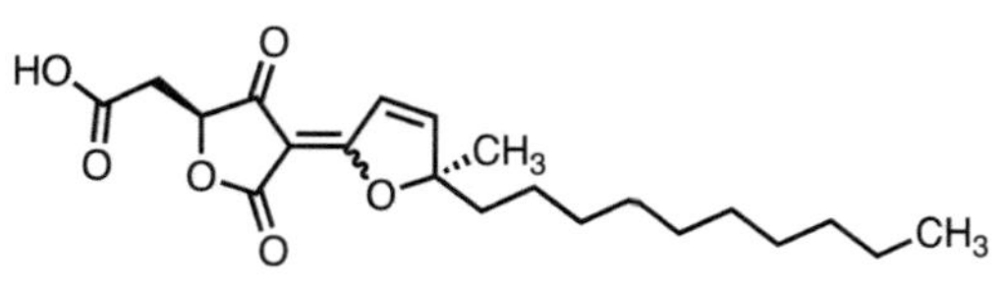

Pestalotic acid A (**2**)

Figure 3. Structure of 4,6′-anhydrooxysporidinone (**1**) and pestalotic acid A (**2**).

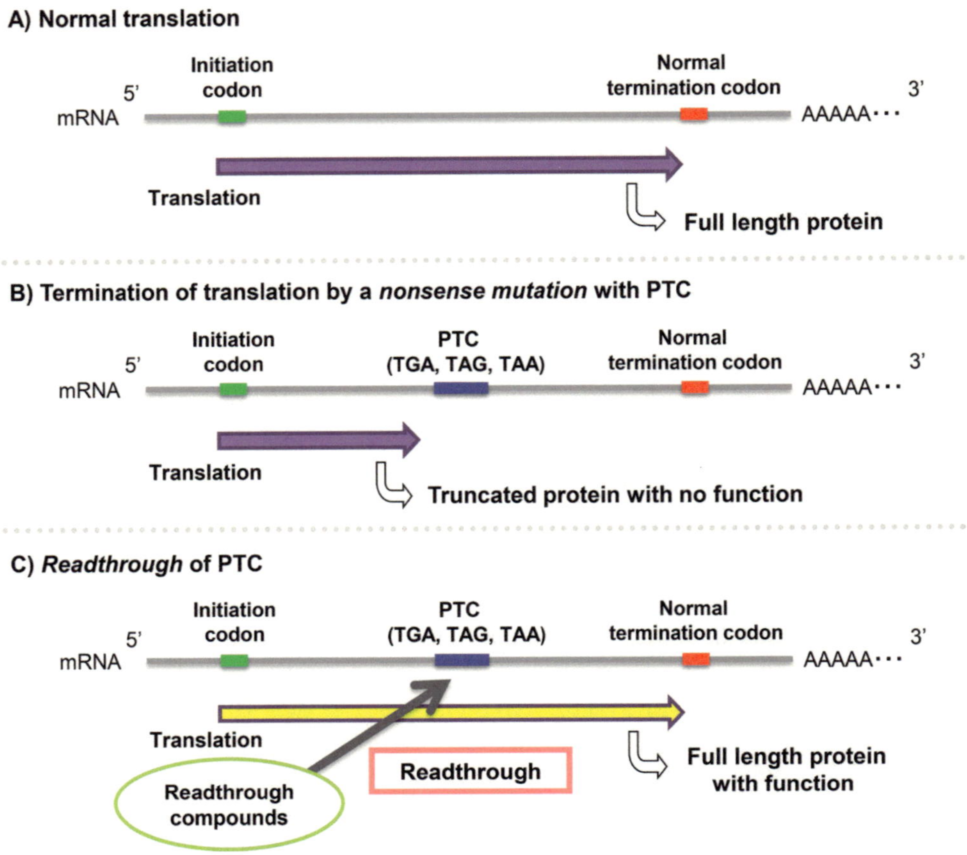

Figure 4. Nonsense mutation as a premature termination codon (PTC) and readthrough compounds. (A) mRNAs containing no PTC are translated into full-length and functional proteins. (B) In the case of mRNAs containing PTC, translation stops at PTC and non-functional truncated proteins are synthesized. (C) In the presence of readthrough compounds, even mRNAs containing PTC are translated into full-length and functional proteins.

structure [66], has readthrough activity and restores dystrophin expression in the muscles of *mdx* mice (**Figure 5**) [67]. In our structure-activity relationship study of (+)-negamycin, we discovered several more potent derivatives, including Leucyl-3-*epi*-deoxynegamycin (TCP-126, **7**) and TCP-112 (**8**) (**Figure 5**) [68, 69]. However, the activities of these compounds are not sufficient for medicine, and the mechanism of action of the readthrough activity remains to be elucidated.

To discover novel readthrough compounds, we constructed yeast strains for the screening of readthrough compounds using 12geneΔ0HSR. *ADE2* is an enzyme that is essential to producing adenine in live yeast systems, and its mutation induced the accumulation of red pigment in vacuoles [70]. One of the *ade2* auxotroph markers, *ade2–101*, has a nonsense mutation (*ochre*) at 190 bp [71]. Therefore, we introduced PTCs at the same site as in the *ADE2* gene and inserted the *ADE2* loci of 12geneΔ0HSR by pop-in/pop-out. The resulting strains 12geneΔ0HSR *ade2-E64X* required adenine for growth and formed red colonies in adenine-limited medium (**Figure 6A**). In contrast, most of the colonies appeared white on

http://dx.doi.org/10.5772/intechopen.70664

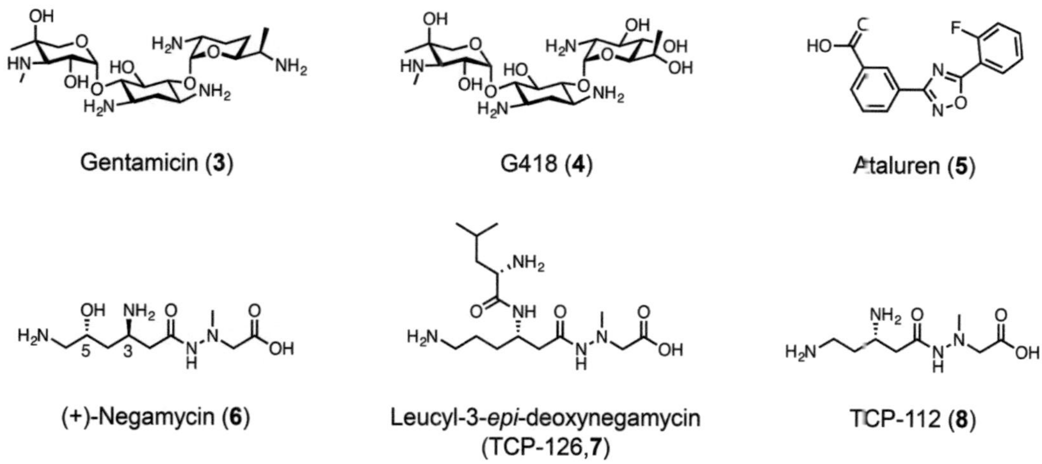

Figure 5. Structure of readthrough compounds. Gentamicin (**3**) and G418 (**4**) are aminoglycoside-type readthrough compounds. Aataluren (**5**), (+)-negamycin (**6**), and negamycin derivatives (Leucyl-3-*epi*-deoxynegamycin (TCP-126, **7**), TCP-112 (**8**)) are non-aminoglycoside-type readthrough compounds.

medium containing TCP-126 (**Figure 6B**), suggesting that TCP-126 evoked readthrough in *ade2-E64X*. In addition, DMSO solution (3 μl) containing readthrough compounds (G418 or negamycin analogues including TCP-126) induced the white halo on the 12geneΔ0HSR *ade2-E64X* strain-inoculated plate after 4 days incubation (**Figure 6C**). These results indicated that 12geneΔ0HSR *ade2-E64X* is suitable for use in the qualitative analysis of readthrough activity.

Next, we initiated a high-throughput screening of the readthrough compounds based on the halo assay using chemical library. This screening is underway, but already several hit compounds have been found, including rapamycin (**9**) [72], wortmannin (**10**) [72], and A23187 (**11**) [73] (**Figure 7**). These data provided further evidence of the usefulness of the 12geneΔ0HSR *ade2-E64X* strains for identifying and elucidating the mechanism of action of readthrough drugs.

2.3. Application 2: target ID

Since our strains show multidrug sensitivity without a decrease in genetic availability, they should also be useful for performing target ID for drugs and the mechanism evaluation of compounds, especially those which are only available in limited amounts, such as natural products. Here we show an example of target ID [50]. Eudistomin C (EudiC, **Figure 8**), a natural product isolated from the Caribbean tunicate *Eudistoma olivaceum* [74, 75] shows broad-spectrum antiviral activity [76]. Because of a unique structural feature, oxathiazepine ring attached to a tetrahydro-β-carboline, EudiC has attracted attention as a lead compound for antiviral medicines. However, several trials for its clinical development have failed due to the strong cytotoxicity of EudiC. To reveal the cause of the cytotoxicity of EudiC, it is important to identify the target molecule responsible for the cytotoxicity of EudiC. By using the yeast genetic approach, we found that a mutation in the RPS14A gene confers EudiC-specific

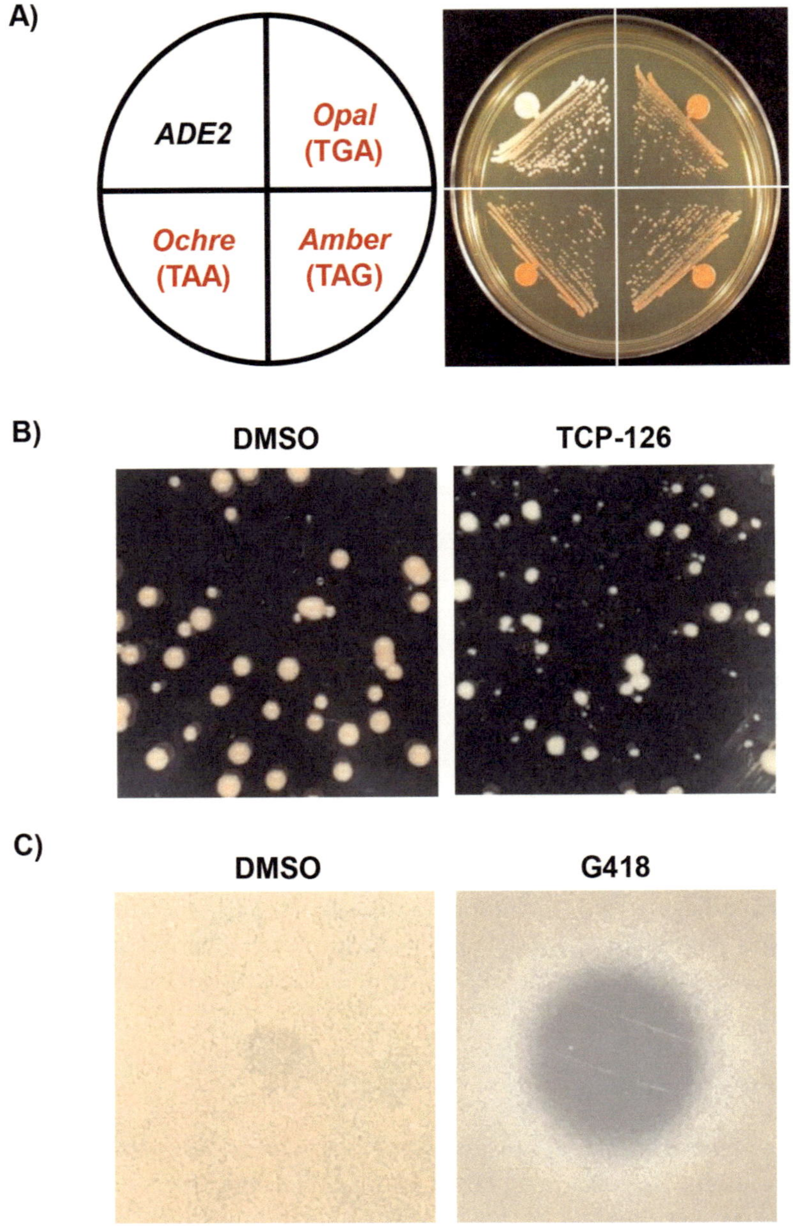

Figure 6. The color of 12geneΔ0HSR *ade2-E64X* strains turned from red to white in the presence of readthrough compounds. (A) 12geneΔ0HSR *ade2-E64X* strains were plated on YPD containing 0.0005% adenine for 4 days. The wild-type strain (12geneΔ0HSR) formed white colonies, but 12geneΔ0HSR *ade2-E64X* strains formed red colonies. (B) 12geneΔ0HSR *ade2-E64X* strains were plated on SC-ADE + 0.0045% adenine with or without luecyl-3-*epi*-deoxy-negamycin (TCP-126) for 4 days. The colonies formed on medium containing TCP-126 were white, suggesting that TCP-126 evoked readthrough activity in the 12geneΔ0HSR *ade2-E64X (TGA)* strain. (C) DMSO and G418 were spotted on 0.5% agar containing 12geneΔ0HSR *ade2-E64X* strains overlaid on YPD containing 0.0005% adenine. After 4 days incubation, the halo that formed around the G418 was white.

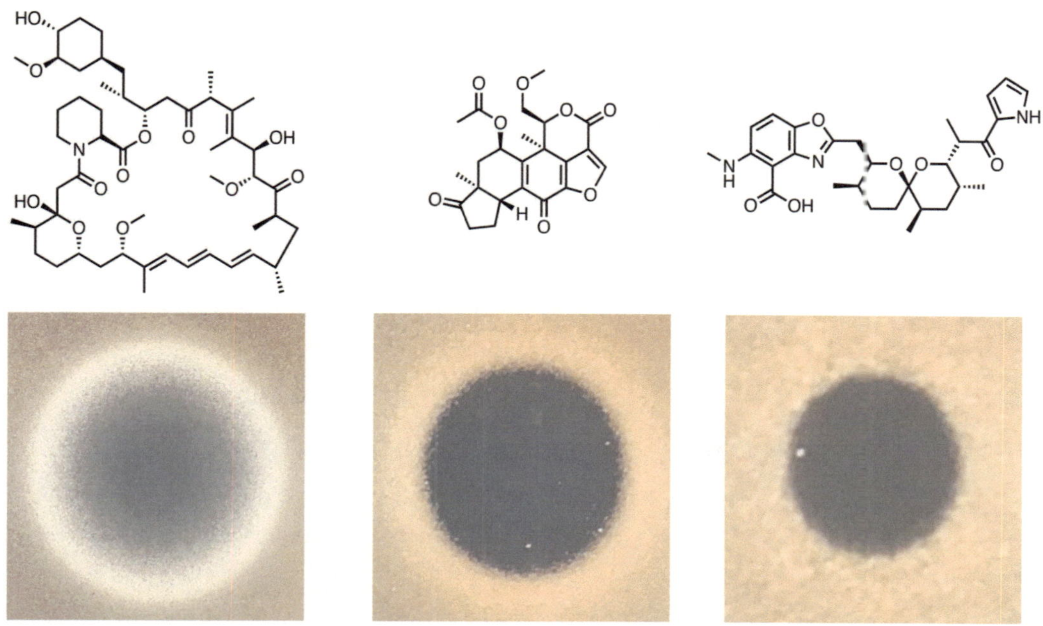

Rapamycin (**9**) Wortmannin (**10**) A23187 (**11**)

Figure 7. Compounds showing readthrough activities in our screening. Rapamycin (**9**), wortmannin (**10**), and A23187 (**11**) were found as readthrough compounds in our assay system. The structures and haloes of these compounds are shown.

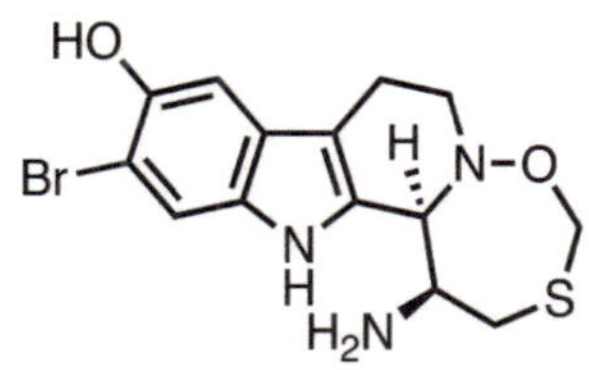

Figure 8. Chemical structure of eudistomin C (EudiC).

resistance [50]. The work flow is shown in **Figure 9**. We used dTC033, one of the multidrug-sensitive yeast strains which lacks 12 genes of the drug-efflux system. The sensitivity of dTC033 against EudiC was 25-fold higher than that of the parental strain BY4741. We isolated the 59 spontaneous mutants that show EudiC resistance. We then crossed these 59 EudiC-resistant strains with OTA014, which has the same genotype as dTC033 (except for the mating type and *RME1(ins-308A)* mutation), and confirmed that 34 of the strains showed dominant resistance. Dominant resistance is predicted to be the mutation in target molecules which inhibits drug-target interaction rather than a lack of cell death signals activated by EudiC treatment (**Figure 9**). These 34 strains were further tested for their EudiC resistance under a higher concentration of EudiC, and 11 strains were selected as strongly resistant mutants. To confirm that the EudiC-resistant mutations of these mutants were not related to multi-drug-resistance mechanisms such as drug efflux pump up-regulation, we checked the sensitivity

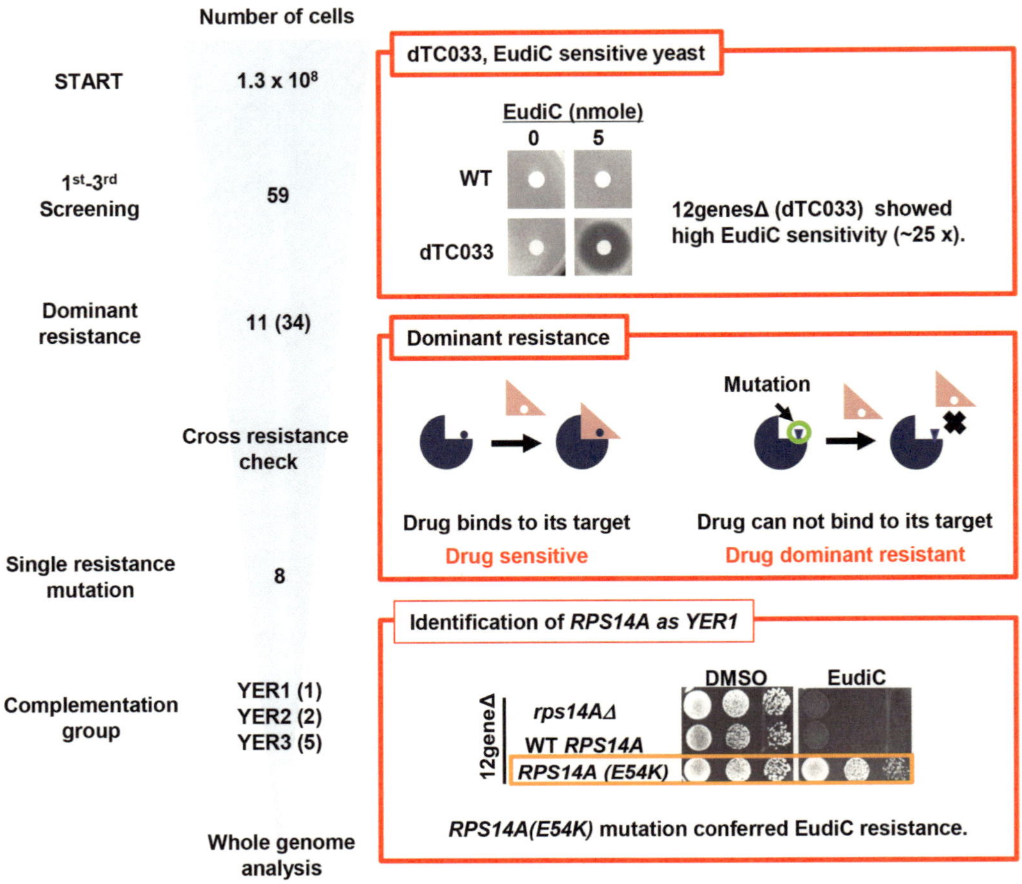

Figure 9. The work flow of the identification of RPS14A as a target of EudiC.

of these mutants against several compounds. These strains did not show cross-resistance against 4-nitroquinoline 1-oxide, digitonin, cycloheximide, or rhodamine 6G, suggesting that these mutants obtained specific resistance against EudiC. We speculate that our strain cannot obtain cross-resistance easily due to its lacking all ABC transporters on the plasma membrane. To select mutants which have a single mutation responsible for EudiC resistance, we performed a tetrad analysis of the spores derived from the diploid of the 11 selected mutants and confirmed that 8 of the strains showed a 2:2 segregation pattern for EudiC resistance. These eight strains were classified into three complementation groups, which we named YER1 (1 strain), YER2 (2 strains), and YER3 (5 strains). "YER" stands for Yeast Eudistomin C Resistance. Whole-genome sequence analysis of the YER strains and further confirmatory analyses, including the disruption of mutated genes in YER strains and the re-introduction of identified mutations into wild-type strains (**Figure 9**), revealed that YER1 is *RPS14A(E54K)*. Unless we checked all of the gene mutations found in the coding region, we failed to identify the mutations in YER2 and YER3, suggesting that the YER2 and YER3 mutations were located on the noncoding region or repetitive sequences—for example, rDNA. *RPS14A* encodes a component of the 40S ribosome, uS11, which participates not only in protein translation but also in 18S ribosomal RNA (rRNA) maturation (20S to 18S processing) in ribosome biogenesis with Fap7p [77]. To distinguish the effect of EudiC on uS11, we performed biochemical analysis using biotinylated EudiC and purified ribosome complexes. Because biotinylated EudiC failed to pull Fap7p down and no effect on 18S maturation processes was observed, it was confirmed that EudiC targets the matured 40S ribosome and inhibits protein translation but not rRNA maturation [50].

Collectively, our target ID studies of EudiC suggested the mode of action of EudiC cytotoxicity and indicated that our sensitive strains would be quite useful for performing drug target IDs in a relatively short period.

3. Conclusions and perspective

In the field of chemical biology, several model organisms, including yeast, worms, flies, and mice, have been used. Yeast is one of the most-used model organisms due to its ease of handling and its genetic availability, but its drug resistance is sometimes an obstacle to investigation. To overcome this problem, we constructed two multidrug-sensitive yeast strains, 12geneΔ0HSR and 12geneΔ0HSR-iERG6. These strains not only show a broad spectrum of drug sensitivities against compounds for which resistance is shown by both ABC transporters and ergosterol without influencing transformation, mating, or sporulation efficiency, but they are also useful for drug screening. Indeed, we performed a screening of antifungal compounds and protein translation regulators which skip stop codons and found some promising candidates. Using 12geneΔ0HSR-iERG6, we succeeded in improving the hit rate of drug screening from microbial broth. The screening of microbial broth which inhibits the growth of 12geneΔ0HSR-iERG6 but not of the quadruplex mutant identified novel compounds suggested that our multidrug-sensitive strain-based screening using previously tested chemical sources in yeast screening could identify new bioactive compounds. Furthermore, as our screening system for readthrough compounds, genetically modified multidrug-sensitive strains can be applied for several types of screening such as a yeast 2-hybrid system-based protein-protein interaction modulators screening. Recently, a yeast 3-hybrid system has been applied for drug-protein interaction analysis [78]. In this study, the *pdr5 snq2 yor1* triple mutant was used to increase the sensitivity of the system [78]. Our multidrug-sensitive yeast strain was thus shown to be useful for this kind of analysis. Moreover, we expect that the 12geneΔ0HSR and 12geneΔ0HSR-iERG6 strains will also be useful tools for genome-wide chemical biology studies such as synthetic lethal/sick genetic interaction analyses [19, 20], genome-wide overexpression screening [21], and haploinsufficiency-chemical sensitive assays [22]. In addition, the genetic approach using our strains identified the 40S ribosome component uS11 as a target molecule of the cytotoxicity caused by the antiviral compound EudiC. Because it has been reported that protein translation is one of the targets for antiviral agents [79–81], the effect on the 40S ribosome and the inhibition of translation by EudiC may cause both the cytotoxicity and the antiviral activity. In contrast, it has also been reported that the uS11 protein interacts with the eS1 and eS26 proteins, which form part of the mRNA exit tunnel [82], and that the eS1 protein is one of the contact sites for hepatitis C virus internal ribosome entry sites (IRES) [83, 84]. These reports might suggest that EudiC decreases the interaction between ribosomes and some of the viral IRES, and efficiently inhibits the translation of viral proteins compared to that of host mRNA. Elucidating the detailed inhibitory mechanism of EudiC on protein translation and its effects on IRES-dependent translation might promote the development of EudiC as a novel antiviral medicine.

Recently, it has been reported that RNAseq combined with Crisper/Cas9-based genome-editing technologies is useful for target ID in mammalian cells [25]. Identification of the drug target using our multidrug-sensitive strains and confirmation of the identified mutation in mammalian cells by Crisper/Cas9-based genome editing will reveal the mechanisms of drugs in more detail. Our multidrug-sensitive strains have the potential to facilitate chemical genetic studies and contribute to the development of medicines in the future.

Author details

Takumi Chinen[1], Keisuke Hamada[2], Akihiro Taguchi[2], Yukihiro Asami[3,4], Kazuro Shiomi[3,4], Yoshio Hayashi[2] and Takeo Usui[5]*

*Address all correspondence to: usui.takeo.kb@u.tsukuba.ac.jp

1 Department of Molecular Genetics, Division of Centrosome Biology, National Institute of Genetics, Mishima, Shizuoka, Japan

2 Department of Medicinal Chemistry, School of Pharmacy, Tokyo University of Pharmacy and Life Sciences, Hachioji, Tokyo, Japan

3 Graduate School of Infection Control Sciences, Kitasato University, Minato-ku, Tokyo, Japan

4 Kitasato Institute for Life Sciences, Kitasato University, Minato-ku, Tokyo, Japan

5 University of Tsukuba, Tsukuba, Japan

References

[1] Katz L, Baltz RH. Natural product discovery: Past, present, and future. Journal of Industrial Microbiology & Biotechnology. 2016;**43**(2):155-176. DOI: 10.1007/s10295-015-1723-5

[2] Towle MJ, Salvato KA, Budrow J, Wels BF, Kuznetsov G, Aalfs KK, et al. In vitro and in vivo anticancer activities of synthetic macrocyclic ketone analogues of halichondrin B. Cancer Research. 2001;**61**(3):1013-1021

[3] Nastrucci C, Cesario A, Russo P. Anticancer drug discovery from the marine environment. Recent Patents of Anticancer Drug Discovery. 2012;**7**:218-232

[4] Chiba H, Tagami K. Research and development of HALAVEN (Eribulin Mesylate). Journal of Synthetic Organic Chemistry, Japan. 2011;**69**(5):600-610. DOI: 10.5059/yukigoseikyokaishi.69.600

[5] Low WK, Dang Y, Schneider-Poetsch T, Shi Z, Choi NS, Merrick WC, et al. Inhibition of eukaryotic translation initiation by the marine natural product pateamine a. Molecular Cell. 2005;**20**(5):709-722. DOI: 10.1016/j.molcel.2005.10.008

[6] Feling RH, Buchanan GO, Mincer TJ, Kauffman CA, Jensen PR, Fenical W. Salinosporamide A: A highly cytotoxic proteasome inhibitor from a novel microbial source, a marine bacterium of the new genus Salinospora. Angewandte Chemie International Edition. 2003;**42**(3):355-357. DOI: 10.1002/anie.200390115

[7] Ueda H, Nakajima H, Hori Y, Fujita T, Nishimura M, Goto T, et al. FR901228, a novel antitumor bicyclic depsipeptide produced by Chromobacterium violaceum no. 968. I. Taxonomy, fermentation, isolation, physico-chemical and biological properties, and antitumor activity. Journal of Anitibiotics (Tokyo). 1994;**47**(3):301-310

[8] Yoshida M, Beppu T. Reversible arrest of proliferation of rat 3Y1 fibroblasts in both the G1 and G2. Experimental Cell Research. 1988;**177**(1):122-131

[9] Kino T, Hatanaka H, Hashimoto M, Nishiyama M, Goto T, Okuhara M, et al. FK-506, a novel immunosuppressant isolated from a Streptomyces. I. Fermentation, isolation, and physico-chemical and biological characteristics. Journal of Antibiotics (Tokyo). 1987;**40**(9):1249-1255

[10] Ingber D, Fujita T, Kishimoto S, Sudo K, Kanamaru T, Brem H, et al. Synthetic analogues of fumagillin that inhibit angiogenesis and suppress tumour growth. Nature. 1990;**348**(6301):555-557. DOI: 10.1038/348555a0

[11] Steegmaier M, Hoffmann M, Baum A, Lénárt P, Petronczki M, Krssák M, et al. BI 2536, a potent and selective inhibitor of polo-like kinase 1, inhibits tumor growth in vivo. Current Biology. 2007;**17**(4):316-322. DOI: 10.1016/j.cub.2006.12.037

[12] Chinen T, Liu P, Shioda S, Pagel J, Cerikan B, Lin TC, et al. The γ-tubulin-specific inhibitor gatastatin reveals temporal requirements of microtubule nucleation during the cell cycle. Nature Communications. 2015;**6**:8722. DOI: 10.1038/ncomms9722

[13] Schenone M, Dančík V, Wagner BK, Clemons PA. Target identification and mechanism of action in chemical biology and drug discovery. Nature Chemical Biology. 2013;**9**(4):232-240. DOI: 10.1038/nchembio.1199

[14] Laffitte E, Revuz J. Thalidomide: An old drug with new clinical applications. Expert Opinion on Drug Safety. 2004;**3**(1):47-56. DOI: 10.1517/14740338.3.1.47

[15] Ito T, Handa H. Cereblon and its downstream substrates as molecular targets of immunomodulatory drugs. International Journal of Hematology. 2016;**104**(3):293-299. DOI: 10.1007/s12185-016-2073-4

[16] Ito T, Ando H, Suzuki T, Ogura T, Hotta K, Imamura Y, et al. Identification of a primary target of thalidomide teratogenicity. Science. 2010;**327**(5971):1345-1350. DOI: 10.1126/science.1177319

[17] Futamura Y, Kawatani M, Kazami S, Tanaka K, Muroi M, Shimizu T, et al. Morphobase, an encyclopedic cell morphology database, and its use for drug target identification. Chemistry & Biology. 2012;**19**(12):1620-1630. DOI: 10.1016/j.chembiol.2012.10.014

[18] Muroi M, Kazami S, Noda K, Kondo H, Takayama H, Kawatani M, et al. Application of proteomic profiling based on 2D-DIGE for classification of compounds according to the mechanism of action. Chemistry & Biology. 2010;**17**(5):460-470. DOI: 10.1016/j.chembiol.2010.03.016

[19] Tong AH, Evangelista M, Parsons AB, Xu H, Bader GD, Pagé N, et al. Systematic genetic analysis with ordered arrays of yeast deletion mutants. Science. 2001;**294**(5550):2364-2368. DOI: 10.1126/science.1065810

[20] Parsons AB, Brost RL, Ding H, Li Z, Zhang C, Sheikh B, et al. Integration of chemical-genetic and genetic interaction data links bioactive compounds to cellular target pathways. Nature Biotechnology. 2004;**22**(1):62-69. DOI: 10.1038/nbt919

[21] Nishimura S, Arita Y, Honda M, Iwamoto K, Matsuyama A, Shirai A, et al. Marine antifungal theonellamides target 3β-hydroxysterol to activate Rho1 signaling. Nature Chemical Biology. 2010;**6**(7):519-526. DOI: 10.1038/nchembio.387

[22] Giaever G, Shoemaker DD, Jones TW, Liang H, Winzeler EA, Astromoff A, et al. Genomic profiling of drug sensitivities via induced haploinsufficiency. Nature Genetics. 1999;**21**(3):278-283. DOI: 10.1038/6791

[23] Harding MW, Galat A, Uehling DE, Schreiber SL. A receptor for the immunosuppressant FK506 is a cis-trans peptidyl-prolyl isomerase. Nature. 1989;**341**(6244):758-760. DOI: 10.1038/341758a0

[24] Luo L, Parrish CA, Nevins N, McNulty DE, Chaudhari AM, Carson JD, et al. ATP-competitive inhibitors of the mitotic kinesin KSP that function via an allosteric mechanism. Nature Chemical Biology. 2007;**3**(11):722-726. DOI: 10.1038/nchembio.2007.34

[25] Kasap C, Elemento O, Kapoor TM. DrugTargetSeqR: A genomics- and CRISPR-Cas9-based method to analyze drug targets. Nature Chemical Biology. 2014;**10**(8):626-628. DOI: 10.1038/nchembio.1551

[26] Wu CY, Feng Y, Cardenas ER, Williams N, Floreancig PE, De Brabander JK, et al. Studies toward the unique pederin family member psymberin: Structure-activity relationships, biochemical studies, and genetics identify the mode-of-action of psymberin. Journal of the American Chemical Society. 2012;**134**(46):18998-19003. DOI: 10.1021/ja 3057002

[27] Miyamoto Y, Machida K, Mizunuma M, Emoto Y, Sato N, Miyahara K, et al. Identification of *Saccharomyces cerevisiae* isoleucyl-tRNA synthetase as a target of the G1-specific inhibitor Reveromycin A. Journal of Biological Chemistry. 2002;**277**(32):28810-28814

[28] Kobayashi Y, Mizunuma M, Osada H, Obayashi YK, Izunuma MM, Sada HO, et al. Identification of *Saccharomyces cerevisiae* ribosomal protein L3 as a target of curvularol, a G1-specific inhibitor of mammalian cells. Bioscience Biotechnology and Biochemistry. 2006;**70**(10):2451-2459

[29] Heitman J, Movva NR, Hall MN. Targets for cell cycle arrest by the immunosuppressant rapamycin in yeast. Science. 1991;**253**(5022):905-909

[30] Schreiber SL. The immunophilins their immunosuppressive ligands molecular recognition by the immunophilins. Science. 1991;**251**(4991):283-287

[31] Li R, Murray AW. Feedback control of mitosis in budding yeast. Cell. 1991;**66**(3):519-531

[32] Hoyt MA, Totis L, Roberts BT. *S. cerevisiae* genes required for cell cycle arrest in response to loss of microtubule function. Cell. 1991;**66**(3):507-517

[33] Kelling J, Sullivan K, Wilson L, Jordan MA. Suppression of centromere dynamics by Taxol in living osteosarcoma cells. Cancer Research. 2003;**63**(11):2794-2801. DOI: 10.1242/jcs.024018

[34] Groen AC, Needleman D, Brangwynne C, Gradinaru C, Fowler B, Mazitschek R, et al. A novel small-molecule inhibitor reveals a possible role of kinesin-5 in an astral spindle-pole assembly. Journal of Cell Science. 2008;**121**(14):2293-2300. DOI: 10.1242/jcs.024018

[35] Kapoor TM, Mayer TU, Coughlin ML, Mitchison TJ. Probing spindle assembly mechanisms with monastrol, a small molecule inhibitor of the mitotic kinesin, Eg5. Journal of Cell Biology. 2000;**150**(5):975-988

[36] Kesisova IA, Nakos KC, Tsolou A, Angelis D, Lewis J, Chatzaki A, et al. Tripolin A, a novel small-molecule inhibitor of aurora A kinase, reveals new regulation of HURP's distribution on microtubules. PLoS One. 2013;**8**(3):e58485. DOI: 10.1371/journal.pone.0058485

[37] J-M W, Chen C-T, Coumar MS, Lin W-H, Chen Z-J, Hsu JT, et al. Aurora kinase inhibitors reveal mechanisms of HURP in nucleation of centrosomal and kinetochore microtubules. Proceedings of the National Academy of Sciences of the United States of America. 2013;**110**(19):E1779-E1787. DOI: 10.1073/pnas.1220523110

[38] Royou A, McCusker D, Kellogg DR, Sullivan W. Grapes (Chk1) prevents nuclear CDK1 activation by delaying cyclin B nuclear accumulation. Journal of Cell Biology. 2008;**183**(1):63-75. DOI: 10.1083/jcb.200801153

[39] Burkard ME, Randall CL, Larochelle S, Zhang C, Shokat KM, Fisher RP, et al. Chemical genetics reveals the requirement for Polo-like kinase 1 activity in positioning RhoA and triggering cytokinesis in human cells. Proceedings of the National Academy of Sciences of the United States of America. 2007;**104**(11):4383-4388. DOI: 10.1073/pnas.0701140104

[40] Lénárt P, Petronczki M, Steegmaier M, Di Fiore B, Lipp JJ, Hoffmann M, et al. The small-molecule inhibitor BI 2536 reveals novel insights into mitotic roles of Polo-like kinase 1. Current Biology. 2007;**17**(4):304-315. DOI: 10.1016/j.cub.2006.12.046

[41] Hewitt L, Tighe A, Santaguida S, White AM, Jones CD, Musacchio A, et al. Sustained Mps1 activity is required in mitosis to recruit O-Mad2 to the Mad1-C-Mad2 core complex. Journal of Cell Biology. 2010;**190**(1):25-34. DOI: 10.1083/jcb.201002133

[42] Santaguida S, Tighe A, D'Alise AM, Taylor SS, Musacchio A. Dissecting the role of MPS1 in chromosome biorientation and the spindle checkpoint through the small molecule inhibitor reversine. Journal of Cell Biology. 2010;**190**(1):73-87. DOI: 10.1083/jcb.201001036

[43] Simon J, Bedalov A. Yeast as a model system for anticancer drug discovery. Nature Reviews Cancer. 2004;**4**(6):481-492. DOI: 10.1038/nrc1372

[44] Hirao M, Posakony J, Nelson M, Hruby H, Jung M, Simon JA, et al. Identification of selective inhibitors of NAD+-dependent deacetylases using phenotypic screens in yeast. Journal of Biological Chemistry. 2003;**278**(52):52773-52782. DOI: 10.1074/jbc.M308966200

[45] Decottignies A, Goffeau A. Complete inventory of the yeast ABC proteins. Nature Genetics. 1997;**15**(2):137-145. DOI: 10.1038/ng0297-137

[46] Bauer BE, Wolfger H, Kuchler K. Inventory and function of yeast ABC proteins: About sex, stress, pleiotropic drug and heavy metal resistance. Biochimica et Biophysica Acta. 1999;**1461**(2):217-236

[47] Jungwirth H, Kuchler K, Yeast ABC. Transporters—A tale of sex, stress, drugs and aging. FEBS Letters. 2006;**580**:1131-1138. DOI: 10.1016/j.febslet.2005.12.050

[48] Chinen T, Ota Y, Nagumo Y, Masumoto H, Usui T. Construction of multidrug-sensitive yeast with high sporulation efficiency. Bioscience Biotechnology and Biochemistry. 2011;**75**(8):1588-1593. DOI: 10.1271/bbb.110311

[49] Chinen T, Nagumo Y, Usui T. Construction of a genetic analysis-available multidrug sensitive yeast strain by disruption of the drug efflux system and conditional repression of the membrane barrier system. The Journal of General and Applied Microbiology. 2014;**60**(4):160-162

[50] Ota Y, Chinen T, Yoshida K, Kudo S, Nagumo Y, Shiwa Y, et al. Eudistomin C, an antitumor and antiviral natural product, targets 40S ribosome and inhibits protein translation. ChemBioChem. 2016;**17**:1616-1620. DOI: 10.1002/cbic.201600075

[51] Bedalov A, Gatbonton T, Irvine WP, Gottschling DE, Simon JA. Identification of a small molecule inhibitor of Sir2p. Proceedings of the National Academy of Sciences of the United States of America. 2001;**98**(26):15113-15118. DOI: 10.1073/pnas.261574398

[52] Storici F, Lewis LK, Resnick MA. In vivo site-directed mutagenesis using oligonucleotides. Nature Biotechnology. 2001;**19**(8):773-776. DOI: 10.1038/90837

[53] Deutschbauer AM, Davis RW. Quantitative trait loci mapped to single-nucleotide resolution in yeast. Nature Genetics. 2005;**37**(12):1333-1340. DOI: 10.1038/ng1674

[54] Dimitrov LN, Brem RB, Kruglyak L, Gottschling DE. Polymorphisms in multiple genes contribute to the spontaneous mitochondrial genome instability of *Saccharomyces cerevisiae* S288C strains. Genetics. 2009;**183**(1):365-383. DOI: 10.1534/genetics.109.104497

[55] Watanabe Y, Suga T, Narusawa S, Iwatsuki M, Nonaka K, Nakashima T, et al. Decatamariic acid, a new mitochondrial respiration inhibitor discovered by pesticidal screening using drug-sensitive *Saccharomyces cerevisiae*. Journal of Antibiotics (Tokyo). 2017;**70**(4):395-399. DOI: 10.1038/ja.2016.164

[56] Wijeratne EMK, Gunatilaka AAL. Biomimetic conversion of (−)-fusoxypyridone and (−)-oxysporidinone to (−)-sambutoxin: Further evidence for the structure of the tricyclic pyridone alkaloid, (−)-fusoxypyridone. Bioorganic & Medicinal Chemistry Letters. 2011;**21**(8):2327-2329. DOI: 10.1016/j.bmcl.2011.02.091

[57] Zhan J, Burns AM, Liu MX, Faeth SH, Gunatilaka AAL. Search for cell motility and angiogenesis inhibitors with potential anticancer activity: Beauvericin and other constituents of two endophytic strains of *Fusarium oxysporum*. Journal of Natural Products. 2007;**70**(2):227-232. DOI: 10.1021/np060394t

[58] Wang Q-X, Li S-F, Zhao F, Dai H-Q, Bao L, Ding R, et al. Chemical constituents from endophytic fungus *Fusarium oxysporum*. Fitoterapia. 2011;**82**(5):777-781. DOI: 10.1016/j.fitote.2011.04.002

[59] Zhang F, Ding G, Li L, Cai X, Si Y, Guo L, et al. Isolation, antimicrobial activity, and absolute configuration of the furylidene tetronic acid core of pestalotic acids A–G. Organic & Biomolecular Chemistry. 2012;**10**(27):5307-5314. DOI: 10.1039/c2ob25469g

[60] Burke JF, Mogg AE. Suppression of a nonsense mutation in mammalian cells in vivo by the aminoglycoside antibiotics G-418 and paromomycin. Nucleic Acids Research. 1985;**13**(17):6265-6272

[61] Barton-Davis ER, Cordier L, Shoturma DI, Leland SE, Sweeney HL. Aminoglycoside antibiotics restore dystrophin function to skeletal muscles of mdx mice. Journal of Clinical Investigation. 1999;**104**(4):375-381. DOI: 10.1172/JCI7866

[62] Sabbavarapu NM, Shavit M, Degani Y, Smolkin B, Belakhov V, Baasov T. Design of novel aminoglycoside derivatives with enhanced suppression of diseases-causing nonsense mutations. ACS Medicinal Chemistry Letters. 2016;**7**(4):418-423. DOI: 10.1021/acsmedchemlett.6b00006

[63] Mingeot-Leclercq MP, Tulkens PM. Aminoglycosides: Nephrotoxicity. Antimicrobial Agents and Chemotherapy. 1999;**43**(5):1003-1012

[64] Hutchin T, Cortopassi G. Proposed molecular and cellular mechanism for aminoglycoside ototoxicity. Antimicrobial Agents and Chemotherapy. 1994;**38**(11) 2517-2520

[65] Welch EM, Barton ER, Zhuo J, Tomizawa Y, Friesen WJ, Trifillis P, et al. PTC124 targets genetic disorders caused by nonsense mutations. Nature. 2007;**447**(7140):87-91. DOI: 10.1038/nature05756

[66] Hamada M, Takeuchi T, Kondo S, Ikeda Y, Naganawa H. A new antibiotic, negamycin. Journal of Antibiotics (Tokyo). 1970;**23**(3):170-171

[67] Arakawa M, Shiozuka M, Nakayama Y, Hara T, Hamada M, Kondo S, et al. Negamycin restores dystrophin expression in skeletal and cardiac muscles of mdx mice. Journal of Biochemistry. 2003;**134**(5):751-758

[68] Taguchi A, Hamada K, Kotake M, Shiozuka M, Nakaminami H, Pillaiyar T, et al. Discovery of natural products possessing selective eukaryotic readthrough activity: 3-epi-deoxynegamycin and its leucine adduct. ChemMedChem. 2014;**9**(10):2233-2237. DOI: 10.1002/cmdc.201402208

[69] Hamada K, Taguchi A, Kotake M, Aita S, Murakami S, Takayama K, et al. Structure-activity relationship studies of 3-epi-deoxynegamycin derivatives as potent readthrough drug candidates. ACS Medicinal Chemistry Letters. 2015;**6**(6):689-694. DOI: 10.1021/acsmedchemlett.5b00121

[70] Roman H. Studies of gene mutation in Saccharomyces. Cold Spring Harbor Symposia on Quantitative Biology. 1956;**21**:175-185

[71] Hieter P, Mann C, Snyder M, Davis RW. Mitotic stability of yeast chromosomes: A colony color assay that measures nondisjunction and chromosome loss. Cell. 1985;**40**(2):381-392

[72] Pal M, Ishigaki Y, Nagy E, Maquat LE. Evidence that phosphorylation of human Upf1 protein varies with intracellular location and is mediated by a wortmannin-sensitive and rapamycin-sensitive PI 3-kinase-related kinase signaling pathway. RNA. 2001;**7**:5-15

[73] Nickless A, Jackson E, Marasa J, Nugent P, Mercer RW, Piwnica-Worms D, et al. Intracellular calcium regulates nonsense-mediated mRNA decay. Nature Medicine. 2014;**20**(8):961-968. DOI: 10.1038/nm.3620

[74] Rinehart KL, Kobayashi J, Harbour GC, Hughes RG Jr, Mizsak SA, Scahill TA. Eudistomins C, E, K, and L, potent antiviral compounds containing a novel oxathiazepine ring from the Caribbean tunicate Eudistoma olivaceum. Journal of the American Chemical Society. 1984;**106**(5):1524-1526. DOI: 10.1021/ja00317a079

[75] Rinehart KL, Kobayashi J, Harbour GC, Gilmore J, Mascal M, Holt TG, et al. Eudistomins A-Q, β-carbolines from the antiviral Caribbean tunicate *Eudistoma olivaceum*. Journal of the American Chemical Society. 1987;**109**(11):3378-3387. DOI: 10.1021/ja00245a031

[76] Lake RJ, Blunt JW, Munro MHG. Eudistomins from the New Zealand ascidian *Ritterella sigillinoides*. Australian Journal of Chemistry. 1989;**42**(7):1201-1206. DOI: 10.1071/CH9891201

[77] Granneman S, Nandineni MR, Baserga SJ. The putative NTPase Fap7 mediates cytoplasmic 20S pre-rRNA processing through a direct interaction with Rps14. Molecular Cellular Biology. 2005;**25**(23):10352-10364. DOI: 10.1128/MCB.25.23.10352-10364.2005

[78] Chidley C, Haruki H, Pedersen MG, Muller E, Johnsson K. A yeast-based screen reveals that sulfasalazine inhibits tetrahydrobiopterin biosynthesis. Nature Chemical Biology. 2011;**7**(6):375-383. DOI: 10.1038/nchembio.557

[79] Puri M, Kaur I, Perugini MA, Gupta RC. Ribosome-inactivating proteins: Current status and biomedical applications. Drug Discovery Today. 2012;**17**(13-14):774-783. DOI: 10.1016/j.drudis.2012.03.007

[80] Ivashkiv LB, Donlin LT. Regulation of type I interferon responses. Nature Reviews Immunology. 2014;**14**(1):36-49. DOI: 10.1038/nri3581

[81] Diamond MS, Farzan M. The broad-spectrum antiviral functions of IFIT and IFITM proteins. Nature Reviews Immunology. 2012;**13**(1):46-57. DOI: 10.1038/nri3344

[82] Ben-Shem A, Garreau de Loubresse N, Melnikov S, Jenner L, Yusupova GYM. The structure of the eukaryotic ribosome at 3.0 Å resolution. Science. 2011;**334**(6062):1524-1529. DOI: 10.1126/science.1212642

[83] Babaylova E, Graifer D, Malygin A, Stahl J, Shatsky I, Karpova G. Positioning of subdomain IIId and apical loop of domain II of the hepatitis C IRES on the human 40S ribosome. Nucleic Acids Research. 2009;**37**(4):1141-1151. DOI: 10.1093/nar/gkn1026

[84] Boehringer D, Thermann R, Ostareck-Lederer A, Lewis JD, Stark H. Structure of the hepatitis C virus IRES bound to the human 80S ribosome: Remodeling of the HCV IRES. Structure. 2005;**13**(11):1695-1706. DOI: 10.1016/j.str.2005.08.008

Section 3

Reproduction and Secretion of the Yeast

Chapter 4

TOR Signaling in Budding Yeast

Yoshiharu Inoue and Wataru Nomura

Additional information is available at the end of the chapter

http://dx.doi.org/10.5772/intechopen.70784

Abstract

TOR (*T*arget *o*f *R*apamycin) is a Ser/Thr kinase that was originally identified by genetic screening using the budding yeast *Saccharomyces cerevisiae*. The TOR protein forms two structurally and functionally distinct complexes (TOR complex 1, TORC1, and TOR complex 2, TORC2). TORC1 is involved in various cellular activities, such as cell growth, ribosome biogenesis, translation initiation, metabolism, stress response, aging, and autophagy. TORC2 is involved in actin organization, sphingolipid biogenesis, and endocytosis. TORC1 plays a central role in the signaling network in response to stimuli coupled to internal and external nutrient conditions, particularly an amino acid sufficiency. A dimeric complex of Rag GTPases, the activity of which is regulated by the guanine nucleotide-loading status, and some regulator proteins communicating with Rag GTPases are involved in the activation of TORC1 by amino acids. In TORC2 signaling, membrane stress appears to be a cue, in which some proteins associated with respective membrane compartments, such as eisosomes, play a role.

Keywords: TOR (*T*arget of *R*apamycin), small GTPase, signal transduction, protein kinase, *Saccharomyces cerevisiae*

1. Introduction

All heterotrophs must take organic compounds from outside of cells to gain energy for various biological activities. For example, since amino acids are components of proteins, an insufficiency in amino acids has serious effects on cellular functions. The bacterial feedback regulation of amino acid biosynthesis at the enzyme and gene expression levels is a well-known mechanism that controls intracellular amino acid levels. In this feedback-regulatory mechanism, an amino acid functions as a signaling molecule in the closed metabolic loop for the production of respective amino acids.

On the other hand, in higher Eukarya, an insufficiency in amino acids has been linked to various metabolic diseases; therefore, sensing amino acid amounts inside and outside of cells through the transmission of signals needs to be strictly controlled. TOR (*T*arget *o*f *R*apamycin) is one of the nutritional signaling mechanisms that is evolutionarily conserved in eukaryotes from yeast to humans. TOR, a Ser/Thr kinase, is involved in two complexes: TOR complex 1 (TORC1) and TOR complex 2 (TORC2), which are distinctly different structurally and functionally. TORC1 is in an active form when amino acids are abundant, namely conditions under which cellular activities that promote cell growth proceed dynamically. TORC1 signaling activates anabolic processes such as protein/lipid synthesis and ribosome biogenesis, which are linked to cell growth. TORC1 signaling also inhibits catabolic processes including autophagy, a bulk protein degradation system. Therefore, the breakdown of TORC1 signaling in humans has been linked to various diseases including cancers and metabolic disorders [1]. TORC2 is involved in the polarized organization of the actin cytoskeleton, endocytosis, and sphingolipid biosynthesis. Physiological cues to activate TORC2 signaling in mammalian cells are insulin and insulin-like growth factors; however, the mechanisms by which these hormones activate mammalian TORC2 have not yet been elucidated in detail. Furthermore, despite the evolutionary conservation of TORC2 in lower eukaryotes such as yeast, the growth factor-like hormonal-regulatory mechanism for the promotion of cell growth is not conserved in yeast. TORC1 signaling has been extensively examined using rapamycin, a potent inhibitor of TORC1 signaling, whereas TORC2 is insensitive to rapamycin. Therefore, although the mechanisms underlying the amino acid-induced activation of TORC1 signaling have been investigated in detail, limited information is currently available on the activation mechanisms of TORC2, which may be explained, at least in part, by the absence of TORC2-specific inhibitors, such as rapamycin for TORC1. However, many important insights into TORC2 signaling have been provided by the budding yeast *Saccharomyces cerevisiae*, which is an excellent model organism. This chapter overviews TOR signaling in the budding yeast, with a particular focus on the regulatory machinery and cues for the activation of TORC1 and TORC2, and compares it with that in mammalian cells. Since TOR is a master regulator of cell growth, the outputs of TOR signaling also cover a broad range of biological activities. More information on the downstream outputs of TOR signaling in yeast is available in recent reviews and references therein [2–4].

2. TOR: a master regulator for cell growth

2.1. Rapamycin and FKBP12

Rapamycin is a macrolide antifungal chemical that was identified from the bacterium *Streptomyces hygroscopicus*, which was isolated from a soil sample obtained on the Easter Islands, *Rapa Nui* in the local tongue, and, hence, it was named "rapamycin." Despite being an antifungal drug, rapamycin also exerted immunosuppressive effects; therefore, it was subsequently used as an immunosuppressant in organ transplantation [5]. Rapamycin and its derivatives also exerted antitumor effects, indicating their potential in the treatment of certain cancers [6].

The first approach to investigating the mode of action of rapamycin was biochemical. Since rapamycin was found to inhibit the mammalian immune system, molecule(s) with the ability to bind to rapamycin may be involved in the action of this drug as an immunosuppressant.

Rapamycin was shown to bind to a peptidyl-prolyl *cis-trans* isomerase, also known as FKBP12 (FK506-binding protein 12), which is one of the immunophilins [7]. Yeast has served as an excellent model organism of higher eukaryotes. Since the mechanisms underlying some biological events discovered in this unicellular microorganism to date are conserved among Eukarya, the mode of action of rapamycin in yeast is also expected to be evolutionarily conserved. Based on this concept, rapamycin-binding protein was purified from yeast cell extracts, and its amino acid sequence was partially elucidated. A reverse genetic approach was applied to clone the gene encoding the rapamycin-binding protein using a partial amino acid sequence, and the FKBP12 homologous gene *FPR1* (*F*KBP12 ***p****rolin*e ***r****otamase*) was obtained [8]. Fpr1 is a small protein that consists of only 114 amino acids (molecular weight, 12,157). Gene disruption experiments revealed that *FPR1* was dispensable for the growth of yeast cells [8–11]. However, since Fpr1 was a rapamycin-binding protein in yeast, the disruption of *FPR1* conferred resistance to rapamycin [8, 12]. These findings suggested that the formation of an Fpr1-rapamycin complex was involved in the mode of action of rapamycin in yeast, and this mode of action was observed in an immunophilin-immunosuppressant complex in mammalian cells.

2.2. Discovery of TOR

In order to identify the target of the Fpr1-rapamycin complex, genetic screening using *S. cerevisiae* with resistance to rapamycin was conducted, and consequently, three genes, that is, *TOR1*, *TOR2*, and *FPR1*, were identified [12]. As expected, most mutants (258 clones from 277 rapamycin-resistant mutants) contained recessive mutations in *FPR1* [8]. Similarly, deletion of *FPR1* conferred the recessive resistance to rapamycin, and expression of human FKBP12 restored sensitivity to rapamycin [13]. Two novel genes, *TOR1* and *TOR2* [8], which were also referred to as *DRR1* and *DRR2*, respectively, for dominant rapamycin resistance [13], were identified.

TOR1 and *TOR2* encode large-molecular-weight proteins (molecular weight, >280 kDa). The Tor1 (2470 amino acids) and Tor2 (2474 amino acids) proteins share 67% identity at the amino acid-sequence level and were initially considered to be lipid kinases (phosphatidylinositol kinases). However, neither proteins exhibited lipid kinase activity; they were later found to be phosphatidylinositol kinase-related kinases (PIKKs). Mutations occurring in *TOR1* (*TOR1-1*) and *TOR2* (*TOR2-1*) that conferred resistance to rapamycin were identified as a single amino acid substitution, that is, Ser1972Arg in Tor1 and Ser1975Ile in Tor2. In contrast to yeast, which possesses two *TOR* genes, mammalian cells have a single *TOR* (mTOR, mammalian TOR) gene. mTOR was initially designated as mammalian TOR, but has recently been referred to as mechanistic TOR, which includes not only mammalian TOR but also all other TORs, such as yeast Tor1 and Tor2.

Rapamycin itself does not directly bind to the TOR protein, whereas the Fpr1-rapamycin complex binds to the Tor1 or Tor2 protein, thereby inhibiting the protein kinase activity of TOR [14–17]. *TOR1-1* and *TOR2-1* produce Tor1 and Tor2 proteins, respectively, without affinity or with low affinity to the Fpr1-rapamycin complex; therefore, mutants with these alleles develop resistance to rapamycin. Similarly, the FKBP12-rapamycin complex binds to mTOR in order to inhibit its activity in mammalian cells [18].

2.3. Domain structure of TOR

The domain structures and amino acid sequences of all TOR proteins are evolutionarily conserved. Both Yeast Tor1 and Tor2 contain the following domains (in the direction from the N-terminus to the C-terminus): HEAT repeats, FAT, FRB, kinase, FIT, and FATC (**Figure 1**). These domains are also found in the mTOR protein in the same order. Each HEAT motif (originally identified in ***H***untingtin, ***e***longation factor 3, protein phosphatase 2***A*** (PP2A), and ***T***OR) consists of approximately 40 amino acid residues that form anti-parallel alpha-helices, and Tor1/Tor2 proteins contain ~20 tandemly repeated HEAT motifs between their N-terminal and central regions. Tor1 and Tor2 bind with their respective subunits that constitute distinct TOR complexes (see subsequent text) through the HEAT repeats. FAT, FRB, kinase, and FATC domains, which are located on the C-terminal to the HEAT repeats, are commonly found in PIKK family members [19–21]. The FAT domain, which was named to represent the main groups in PIKKs (***F***RAP, ***A***TM, and ***T***RRAP), consists of ~500 amino acid residues. The FRB (***F***KBP-***r***apamycin ***b***inding) domain consists of ~100 amino acid residues, and the Fpr1-rapamycin complex binds to this region. The *TOR1-1* and *TOR2-1* mutations conferring resistance to rapamycin occur within the FRB domain, which demonstrates that the Fpr1-rapamycin complex is a true inhibitor of TOR kinase.

2.4. TOR complexes

Although the primary structures of Tor1 and Tor2 share strong similarities, their cellular functions are distinct [14, 22]. The *TOR1* null mutation is viable, whereas the *TOR2* null mutation is not. Previous studies reported that rapamycin treatments mimicked starvation, indicating that TOR is involved in cell growth control in response to nutrients [23, 24]. The findings of genetic analyses on *TOR1* and *TOR2* suggested that the roles of Tor1 and Tor2 are divided into two aspects, that is, some readouts in which TOR signaling is involved are redundantly regulated by Tor1 and Tor2, whereas some are specifically regulated by Tor2. Rapamycin was found to affect cellular events in which Tor1 and Tor2 functioned redundantly. For example, rapamycin inhibits protein synthesis and ribosome biogenesis, but induces autophagy, which occurs under nutrient-starved conditions, and both Tor1 and Tor2 are involved in these events [18]. Meanwhile, the regulation of actin organization, endocytosis, and sphingolipid biosynthesis is controlled by Tor2. The distinction between Tor1- and Tor2-related readouts is due to differences in the complexes in which Tor1 and Tor2 are involved.

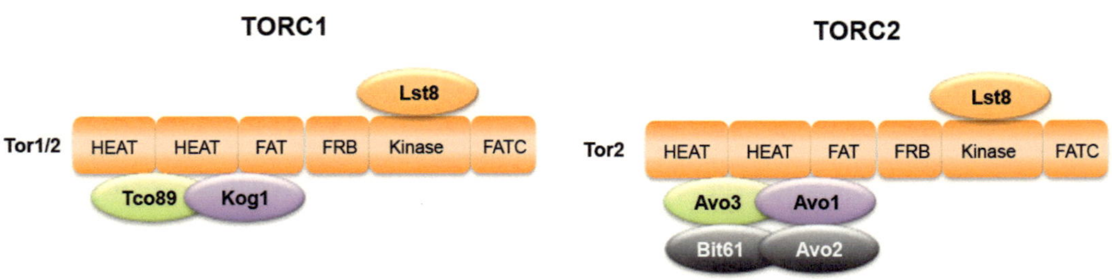

Figure 1. TOR complexes in *S. cerevisiae*.

http://dx.doi.org/10.5772/intechopen.70784

TORC1 involves either Tor1 or Tor2 as the TOR protein, while Kog1, Tco89, and Lst8 are subunits. TORC2 involves Tor2 as the TOR protein with subunits of Avo1, Avo2, Avo3, Bit61, and Lst8. Readouts redundantly regulated by Tor1 and Tor2 are controlled by TORC1, which is sensitive to rapamycin, whereas the specific readouts of Tor2 are regulated by TORC2. TORC2 contains Tor2, in which the FRB domain exists; however, this TOR complex is not sensitive to rapamycin. This issue was resolved using crosslinking-mass spectrometric and electron microscopic analyses, that is, the C-terminal part of Avo3 was close to the FRB domain, which rendered the Fpr1-rapamycin complex incapable of accessing the FRB domain, resulting in TORC2 insensitivity to rapamycin [25].

3. TORC1

3.1. Subunit components

The following components constitute TORC1: Kog1, Tco89, Lst8, and either Tor1 or Tor2 [26–28] (**Figure 1**). Mammalian TORC1 (mTORC1) contains counterparts of each subunit of yeast TORC1, except for Tco89, instead mTORC1 contains PRAS40 (***p***roline-***r***ich ***A***kt ***s***ubstrate of ***40*** kDa) and DEPTOR (***D***isheveled, ***E***gl-10, and ***P***leckstrin domain-containing m***TOR***-interacting protein). mTORC1 forms a dimeric structure [29], and this also appears to be the case for yeast TORC1. The structural integrity of mTORC1 was disrupted by rapamycin [29], whereas all yeast TORC1 components were co-immunoprecipitated by FKBP12 [26], suggesting that rapamycin does not affect the structure of TORC1 in yeast.

Kog1 and mammalian ortholog Raptor (***r***egulatory-***a***ssociated ***p***rotein of m***TOR***) contains the RNC (***R***aptor ***N***-terminal ***c***onserved) domain, through which Kog1/Raptor binds to the TOR protein and the substrates of TORC1. Kog1/Raptor contains three HEAT repeats in the proximity of the C-terminal of the RNC domain and also contains seven WD40 motifs in the C-terminus. Tco89 contains no obvious motifs. Lst8 (mLst8 in mTORC1) contains seven WD40 motifs.

3.2. Activation of TORC1 signaling

3.2.1. *Rag GTPases (Gtr1 and Gtr2)*

When cells are exposed to conditions that are unfavorable for growth, they cease division and remodel cellular metabolism and gene expression profiles to survive under these stressful conditions. The treatment of yeast cells with rapamycin causes multiple phenomena resembling those occurring in cells starved of nutrients, particularly amino acids. Therefore, one of the physiological cues for the activation of TORC1 signaling may be amino acid(s). Upstream module(s) that communicate with TORC1 were revealed by a genetic approach using *S. cerevisiae*. Mutants that are unable to recover cell growth when transferred from nutrient-depleted conditions to nutrient-rich conditions are expected to be defective in module(s) that communicate with TORC1. Since rapamycin mimics amino acid-starved conditions, mutants with the ability to recover from rapamycin-induced growth arrest were screened. *EGO* (***E***xit from rapamycin-induced ***GrO***wth arrest) mutants were identified, in which the Ras-related GTPase (Rag) Gtr2 and the vacuolar membrane-associated proteins Ego1 and Ego3 were included [30]. Another study revealed that modules involved in the trafficking of the general amino acid permease to the cytoplasmic membrane were Gtr1, Gtr2, Ego1, Ego3, and Ltv1 [31].

Gtr1 and Gtr2 belong to the Rag family. Orthologs of Gtr1 and Gtr2 in mammalian cells are RagA/RagB for Gtr1 and RagC/RagD for Gtr2. Amino acid-sequence similarities between RagA and RagB (90% identity) and between RagA/RagB and Gtr1 (48%) are high. This is also the case between RagC and RagD (81%) and between RagC/RagD and Gtr2 (46%). However, amino acid-sequence similarities between RagA/RagB and RagC/RagD and between Gtr1 and Gtr2 are low (approximately <25%) [32–34]. Rag GTPases function as heterodimers that are formed by a combination of one monomer of either RagA or RagB and one monomer of either RagC or RagD [33]. Similarly, Gtr1 and Gtr2 form a heterodimer [34]. Heterodimers with GTP-bound RagA/RagB and GDP-bound RagC/RagD exhibit full activity. This is also the case for *S. cerevisiae* Rag GTPase, that is, $Gtr1^{GTP}$ and $Gtr2^{GDP}$ are a dynamic combination that activate TORC1 in yeast (**Figure 2**).

3.2.2. EGO complex (ragulator)

Small GTPases are generally lipid-linked proteins, and lipid modifications enable these proteins to anchor to biological membranes. However, neither Gtr1 nor Gtr2 is modified by lipids.

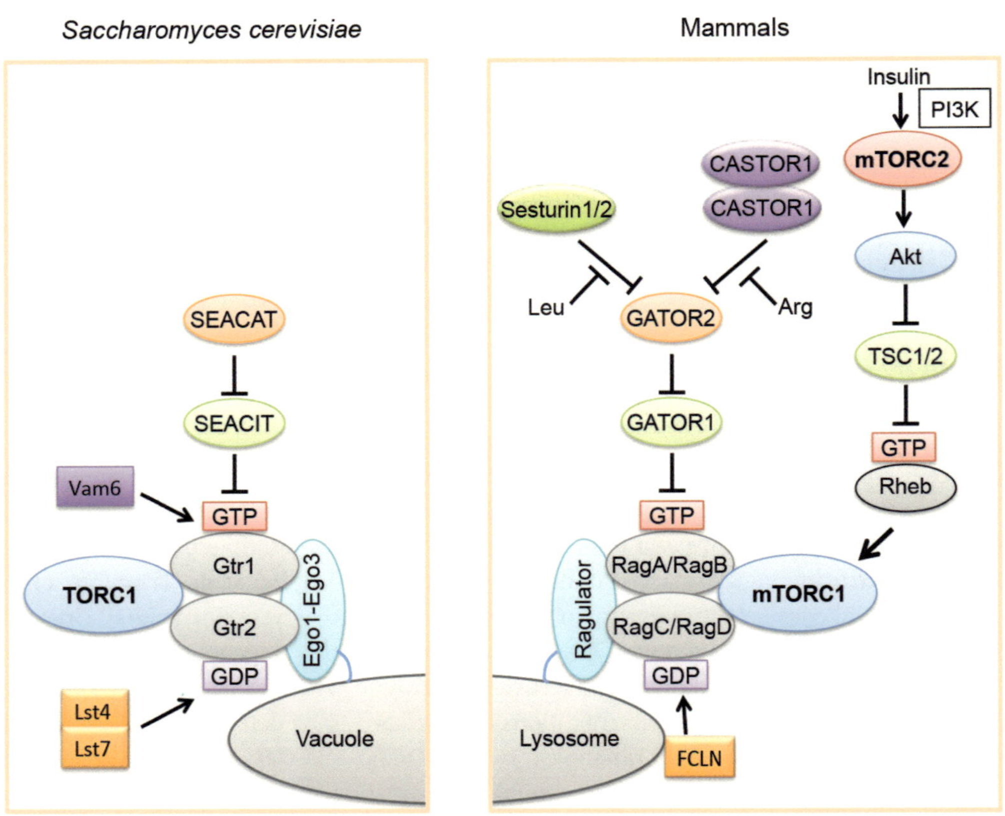

Figure 2. Activation of TORC1 signaling in yeast and mammals.

Ego1, Ego3, and Ego2 were recently found to form the EGO complex, which serves as a scaffold for the Gtr1-Gtr2 heterodimer to anchor to the vacuolar membrane in order to activate TORC1 in response to amino acids in *S. cerevisiae* [35]. Ego1 is a myristoylated and palmitoylated protein that is anchored to the vacuolar membrane through such lipids [36–39]. Ego2 and Ego3 bind to vacuolar membrane-anchored Ego1 [35].

Mammalian cells also contain a large protein complex that functions together with the heterodimeric Rag GTPase, designated "Ragulator" (Rag regulator). Ragulator consists of five subunits, that is, LAMTOR1-5 (LAMTOR, ***L***ysosomal ***A***daptor, and ***M***itogen-activated protein kinase (MAPK), and m***TOR***). Ragulator is anchored to lysosomal membranes through lipid-modified LAMTOR1 [40]; therefore, LAMTOR1 may be a functional homolog of Ego1. LAMTOR 2 and LAMTOR3 form a heterodimer each with a monomer protein, which are structurally and functionally homologous to Ego3 [41]. LAMTOR4 and LAMTOR5 show high structural similarities with Ego2 and Ego4, a paralog of Ego2 [35, 42]. Ragulator and heterodimeric Rag GTPases, which consist of GTP-bound RagA/RgB and GDP-bound RagC/RagD, communicate the signal of an amino acid sufficiency to mTORC1 on lysosomal membranes in mammalian cells (**Figure 2**).

3.2.3. *GEF and GAP for Rag GTPases*

The activities of small GTPases are generally regulated by the status of the guanine nucleotide loaded, which is controlled by the guanine nucleotide exchange factor (GEF) and GTPase-activating protein (GAP). Gtr1-activating factors were screened using a genetic approach, and, as a result, Vam6 (also known as Vps39) was obtained [36]. Vam6/Vps39 exhibited Gtr1 GEF activity *in vitro* [36].

On the other hand, a breakthrough regarding the regulation of Gtr2 was achieved by the discovery that Folliculin (FLCN) tumor suppressor functioned as a positive regulator of RagC/RagD [43, 44]. FLCN forms a complex with either FNIP1 or FNIP2 and has the ability to recruit mTORC1 to lysosomal membranes in response to an amino acid stimulation, thereby activating mTORC1. FLCN-FNIP1/2 complexes are GAPs toward RagC/RagD. A similar mechanism was conserved in yeast, that is, Lst7 and Lst4 are counterparts of FLCN and FNIP1/2, respectively [45]. Lst4 and Lst7 form a stable complex, both of which are necessary for the activation of TORC1 in the presence of amino acids. The Lst4-Lst7 complex preferentially binds to $Gtr2^{GTP}$ in order to enhance the hydrolytic activity of its GTPase activity, thereby yielding $Gtr2^{GDP}$ to activate TORC1 in yeast upon an amino acid stimulation (**Figure 2**).

3.2.4. *SEACIT and SEACAT (GATOR1 and GATOR2)*

What is an upstream regulator(s) of Rag GTPases? In order to solve this question, genome-wide screening was conducted in yeast to discover negative effectors of TORC1 activity, and, as a result, Npr2 and Npr3 were identified [46]. Npr2 and Npr3 form a heterodimer [46]. The coatomer-related Seh1-associated complex (SEAC) that associates with vacuolar membranes was implicated in responses to nitrogen starvation [47–49]. Npr2 and Npr3 together with Iml1/Sea1 form a SEAC subcomplex, which negatively regulates Gtr1 within the EGO complex [50]. A biochemical analysis revealed that Iml1/Sea1 exhibited GAP activity toward Gtr1 *in vitro*.

Npr2 is phosphorylated by an unknown protein kinase and dephosphorylated by PP2A, the phosphorylation status of which correlates with the assembly of this SEAC subcomplex. The Iml1/Sea1-Npr2-Npr3 SAEC subcomplex, which was named SEACIT (for ***SEAC*** subcomplex-***I****nhibiting* ***T****ORC1 signaling*), functions as GAP toward Gtr1, thereby inhibiting TORC1 signaling [50]. The GAP activity of SEACIT is conserved in higher eukaryotes, such as *Drosophila* and humans, that is, DEPDC5-NPRL2-NPRL3 corresponds to the yeast Iml1/Sea1-Npr2-Npr3, and DEPDC5 (ortholog of yeast Iml1/Sea1) directly binds to RagA and enhances the hydrolytic activity of GTP-bound RagA; therefore, the DEPDC5-NPRL2-NPRL3 complex was designated GATOR1 (for ***GAP A****ctivity* ***TO****ward* ***R****ags* ***1***). GATOR1 inactivates mTORC1 in the absence of amino acids (**Figure 2**).

The SEAC of yeast is an octameric complex, that is, SEAC contains Sea2, Sea3, Sea4, Seh1, and Sec13 besides Iml1/Sea1-Npr2-Npr3, which constitutes SEACIT. These proteins constitute the other SEAC subcomplex, which binds to SEACIT in order to inhibit its Gtr1 GAP activity, and, thus, has been designated SEACAT (***SEAC*** subcomplex-***A****ctivating* ***T****ORC1 signaling*) [49]. Orthologs of components in SEACAT also exist in *Drosophila* and mammals, that is, WDR24, WDR59, Mios, Seh1L, and Sec13 in flies and humans, respectively, are Sea2, Sea3, Sea4, Seh1, and Sec13 in yeast, and this complex is referred to as GATOR2. All components in SEACAT and GATOR2 contain beta propeller-forming WD40 motifs, which are characteristic in membrane-coating proteins [51]. Sec13 is a component of COPII, which controls vesicle transport. In addition, Seh1/Seh1L and Sec13 are components of the nuclear pore complex [49].

3.2.5. Upstream modules of GATOR2 (Sestrins and CASTOR)

Amino acid sensors that function upstream of GATOR2 were identified in 2016, that is, Sestrin1/2 as a Leu sensor [52] and CASTOR as an Arg sensor [53]. Previous studies reported that Sestrins interacted with GATOR2 in order to inhibit mTORC1 signaling under amino acid-depleted conditions [54–56]. Wolfson et al. [52] demonstrated that Leu directly bound to Sestrin2 with a dissociation constant of 20 μM, and the binding of Leu to Sestrin2 disrupted the Sestrin2-GATOR2 interaction, thereby enabling GATOR2 to interact with GATOR1. The interaction between GATOR2 and GATOR1 inhibits the GAP activity of GATOR1 toward RagA/RagB, and, consequently, mTORC1 is activated.

The uncharacterized protein CASTOR1 binds to GATOR2, which inhibits GATOR2 binding to GATOR1. CASTOR1 forms a homodimer with CASTOR1 and a heterodimer with CASTOR2, a CASTOR1-related protein. Arginine specifically binds to CASTOR1 with a dissociation constant of ~30 μM, and the binding of Arg to CASTOR1 disrupts the CASTOR1-GATOR2 interaction, which turns CASTOR1 into a homodimer [53]. Liberated GATOR2 interacts with GATOR1 in order to inhibit its GAP activity toward RagA/RagB, which leads to the activation of mTORC1 (**Figure 2**).

Since no orthologs of Sestrins or CASTOR have been found in yeast, the mechanisms by which yeast senses intracellular amino acid availability currently remain unclear. A model in which tRNA functions as a negative regulator of TORC1 kinase activity in yeast was recently proposed [57]. Both amino acid-uncharged and amino acid-charged (aminoacylated) tRNAs inhibited TORC1 kinase activity in an *in vitro* kinase assay. Under nutrition-sufficient conditions, aminoacylated tRNAs predominantly bind to ribosomes for protein synthesis; therefore, tRNAs have fewer opportunities to interact with TORC1 (i.e., TORC1 is active). Upon nutrition starvation, uncharged tRNAs are released from ribosomes and interact with TORC1 in order to inhibit its kinase activity.

Human and yeast cells depleted for Rag GTPase/Gtr remained the ability to respond to amino acid, particularly glutamine [58–60]. It was recently reported that phosphatidylinositol 3-kinase complex Vps34-Vps15, and a vacuolar membrane protein Pib2, which contains a phosphatidylinositol 3-phosphate-binding FYVE (*F*ab1, *Y*OTB, *V*ac1, and *E*EA1) domain played a role in sensing glutamine in the Gtr-independent activation of TORC1 in *S. cerevisiae* [61].

3.3. The TSC1/2-Rheb branch in the activation of mTORC1

In mammalian cells, mTORC1 is activated by another small GTPase Rheb (***R**as **h**omolog **e**nriched in **b**rain*). Similar to other small GTPases, GTP-bound Rheb is a dynamic form in terms of the activation of mTORC1, and the guanine nucleotide status in Rheb is regulated by machinery downstream of growth factor signaling, such as the insulin-signaling pathway. Although the mechanisms by which $Rheb^{GTP}$ stimulate mTORC1 have not yet been elucidated, mTORC1 activity is negatively regulated by the TSC complex, consisting of TSC1, TSC2, and TBC1D7, in which TSC2 functions as GAP toward $Rheb^{GTP}$. Upon a growth factor stimulation, Akt, a member of the AGC kinase family, is activated in a phosphatidylinositol 3-kinase-dependent manner, and activated Akt subsequently phosphorylates TSC2. The TSC complex is localized in the cytoplasm close to lysosomal membranes with which mTORC1 associates via Ragulator, and the phosphorylation of TSC2 alters the localization of the TSC complex away from the lysosome, thereby releasing Rheb from the inhibitory effects induced by the TSC complex [62]. On the other hand, a previous study reported that amino acid deprivation recruited the TSC complex to the lysosome [63], suggesting that the amino acid-dependent activation of mTORC1 is regulated by an interplay between the Rag GTPases-Ragulator branch and the TSC complex-Rheb branch. A recent study reported that Arg is required for the growth factor-dependent delocalization of the TSC complex from the lysosome, which leads to the activation of Rheb, and, thus, mTORC1 [64] (**Figure 2**).

S. cerevisiae does not contain the orthologs of TSC1/2, but has the Rheb homolog, Rhb1; however, there is currently no evidence to show that Rhb1 is a functional homolog of mammalian Rheb [36]. On the other hand, the fission yeast *Schizosaccharomyces pombe* was found to have homologs of TSC1/2 and Rheb [65]. Similar to the mammalian TSC complex, Tsc1 and Tsc2 in fission yeast form a complex, in which Tsc2 functions as a GAP toward Rhb1 GTPase [65–67]. Rhb1 physically interacts with Tor2 (Tor2 in fission yeast corresponds to Tor1 in budding yeast), thereby stimulating TORC1 (Tor2 is involved in TORC1 in fission yeast) activity [65, 68]. Therefore, an *rhb1* mutant showed some phenotypes that are displayed in cells starved of nitrogen [69, 70]. An epistatic analysis showed that Rhb1 functions upstream of Tor2, that is, the activated allele of *tor2*$^{+}$ suppressed the loss of function of *rhb1*$^{+}$ [71].

4. TORC2

4.1. Subunit components

The following components constitute the budding yeast TORC2: Tor2, Avo1, Avo2, Avo3, Bit61, and Lst8 (**Figure 1**). Avo1 has several conserved domains. Avo1 contains an RBD (a ***R****as-****b****inding* ***d****omain*) at the center of its molecule. At the C-terminal region of Avo1, an essential PH (***P****leckstrin* ***h****omology*)-like domain exists, through which TORC2 may tether to the definite region of the plasma membrane called the MCT (***m****embrane* ***c****ompartment-containing* ***T****ORC2*) [72]. The CRIM (***c****onserved* ***r****egion* ***i****n the* ***m****iddle*) domain exists in proximity to the N-terminal side of RBD and has been implicated in binding to the substrates of TORC2 [73, 74]. Avo1 binds to the kinase domain of Tor2 via Lst8 [25].

Avo3 is the largest subunit of TORC2. It functions as a scaffold protein in order to maintain the integrity and function of TORC2 because the loss of Avo3 induced the disassembly of TORC2 [75]. Avo3 contains the ARM (***arm****adillo repeat*)-like domain, which is a similar structure to the HEAT repeats, at the center of its molecule. Repeated ARM units fold together as a superhelical structure to provide a platform to interact with many proteins [76]. Avo3 also interacts with the FAT and kinase domains of Tor2 within TORC2. Avo3 has a RasGEFN domain, which is found in the N-terminal region of GEF proteins toward Ras-like GTPases; however, the function of this domain in Avo3 currently remains unknown. Since the FRB domain of Tor2 within TORC2 is masked by the C-terminal part of Avo3, the accessibility of the Fpr1 (FKBP12)-rapamycin complex to TORC2 is limited, which renders TORC2 insensitive to rapamycin.

Bit61 has a paralog Bit2. Although Bit61 binds to TORC2 through Avo1 and Avo3, it is not an essential subunit for the assembly of TORC2 [25, 75]. The specific functions of Bit61 have not yet been elucidated; however, mammalian orthologs of Bit61 and Bit2 exist (PRR5 also known as Protor-1, and PRR5L also known as Protor-2) and possess an HbrB domain that was found in a fungal *Aspergillus nidulans* protein required for filamentous growth [77].

Avo2 is a yeast TORC2-specific subunit, but is not essential. Avo2 contains ankyrin repeats. Avo2 and Bit61 have been reported to bind to Slm1 and Slm2 proteins, which are involved in the recruitment of Ypk1/Ypk2 to TORC2, thereby phosphorylating them [78].

The core subunits of mammalian TORC2 (mTORC2) include mTOR as the TOR protein, mSin1 (***s****tress-activated protein kinase-****in****teracting protein* ***1***) as the Avo1 ortholog, Rictor (***r****apamycin-****i****nsensitive* ***c****ompanion of m****TOR***) as the Avo3 ortholog, and mLst8 as the Lst8 ortholog. Analogous to the yeast counterpart, mSin1 contains RBD. mSin1 was originally cloned as a factor that interfered with *S. cerevisiae* Ras signaling [79]. mSin1 also contains the CRIM and PH domains, which function in the binding of substrates and tethering to the plasma membrane, respectively.

The ARM-like domain is conserved in Rictor and Avo3, while the RasGEFN domain is not conserved in Rictor. mTORC2 is also insensitive to an acute treatment with rapamycin, the mechanism of which is presumably the same as that elucidated in yeast TORC2. However, in some mammalian cell lines, a prolonged treatment with rapamycin was found to inhibit the interaction between newly synthesized mTOR and Rictor, and mTORC2-Akt signaling was subsequently reduced [80].

4.2. Activation of TORC2

4.2.1. Implication of GTPases

Small GTPase Rag complexes ($RagA^{GTP}/RagB^{GTP}$-$RagC^{GDP}/RagD^{GDP}$ in metazoans, and $Gtr1^{GTP}$-$Gtr2^{GDP}$ in yeast) play pivotal roles in the amino acid-induced activation of TORC1, as described in the previous sections. The other small GTPase Rheb is also involved in the growth factor-mediated activation of mTORC1. Do any small GTPases play roles in the activation of TORC2? In the fission yeast *S. pombe,* genetic screening revealed that the human Rab6 GTPase ortholog Ryh1 was involved in TORC2-Gad8 signaling [81]. *S. pombe* TORC2 phosphorylated Gad8, a member of the AGC kinase family, and a genetic mutation in *ryh1*$^{+}$ markedly decreased the phosphorylation level of Gad8. *sat1*$^{+}$ and *sat4*$^{+}$ genes were predicted to code for GEFs toward Ryh1, and the mutational inactivation of these genes also induced a decrease in the phosphorylation level of Gad8, suggesting that $Rhy1^{GTP}$ is an active form in terms of the activation of TORC2-Gad8 signaling. GTP-locked Rhy1 facilitated the physical interaction between TORC2 and its substrate Gad8. Furthermore, the expression of human Rab6 functionally compensated for the loss of *ryh1*$^{+}$ in *S. pombe* in terms of TORC2 signaling, which implied that Rab GTPase is involved in mTORC2-Akt signaling in mammals, similar to fission yeast. However, since *S. cerevisiae* does not possess the Rab6 ortholog, it currently remains unclear whether this regulatory system is generally conserved in eukaryotes. However, Avo1 and Avo3 contain the RBD and RasGEFN domains, respectively, both of which are related to Ras GTPase; therefore, some small GTPases may be involved in TORC2 signaling in *S. cerevisiae*. Previous studies demonstrated the participation of small GTPases in mTORC2 signaling. Rac1 GTPase was reported to bind directly to mTOR within mTORC1 and mTORC2, which led to the appropriate localization of these TOR complexes to the respective cellular membranes [82]. Rit, a Ras family GTPase, was shown to bind to mTORC2 and subsequently activate it in response to oxidative stress [83]. Since the oxidative stress-responsive activation of TORC2 was also observed in *S. cerevisiae* [84], a similar mechanism by which Ras family GTPase activates TORC2 may be conserved in budding yeast.

4.2.2. Posttranslational modifications in TORC2 components

mTOR is phosphorylated in the growth factor-mediated activation of mammalian TOR signaling. For example, Thr^{2173} in the kinase domain of the mTOR protein is phosphorylated by Akt, which appears to be the negative feedback regulation of mTORC2 signaling. This feedback regulation is also conserved in fission yeast TORC2-Gad8 signaling, that is, Gad8 phosphorylates Thr^{1972} in the ATP-binding domain to reduce Tor1 activity within TORC2 [85]. More than 20 potential phosphorylation sites have been assigned in Rictor [86]. Ser^{260} in the CRIM domain and Thr^{398} in the PH domain in mSin1 are also phosphorylated [87, 88]. A high-throughput phosphoproteomic analysis predicted numerous potential phosphorylation sites in Avo1-3 and Bit61 [89].

Besides phosphorylation, Rictor is known to be acetylated at Lys^{1116}, Lys^{1119}, and Lys^{1125} [90, 91], modifications to which may activate mTORC2 activity.

4.3. Activation of TORC2 signaling

4.3.1. Relationship between membrane tension and the activation of TORC2 signaling in yeast

When TORC2 was observed using GFP-tagged Avo1 or Avo3, its localization was visible as many dots just beneath the plasma membrane. The plasma membrane regions at which patchy TORC2 is located are referred to as the MCT [78]. Although Avo1 contains the PH domain, which has the potential to associate with membrane phospholipids, the underlying mechanisms by which TORC2 localizes to the plasma membrane remain unclear. Other regions on the yeast plasma membrane, referred to as eisosomes, are characterized by their distinctive shape, that is, they are furrows approximately 50-nm deep and 300-nm long on the surface of the plasma membrane [92]. The curvature of the membrane in eisosomes is formed by proteins possessing the BAR (***B***in/***a***mphiphysin/***R***vs) domain, that is, Pil1 and Lsp1. Eisosomes exist in close proximity to the MCT, but never overlap.

Slm1 and its paralog Slm2 are eisosome-residential proteins and are effectors as well as substrates of TORC2. Under normal turgor pressure conditions, Slm1 and Slm2 are predominantly localized in eisosomes; however, following an increase in membrane tension caused by, for example, hypotonic shock or some mechanical stress, Slm1 and Slm2 alter their localization from eisosomes to the MCT and then bind to TORC2 via its components Avo2 and Bit61. Slm1 and Slm2 may recruit Ypk1 to TORC2, and the interaction between TORC2 and its substrate Ypk1 promotes the phosphorylation of Ypk1 (**Figure 3**).

It has not yet been established whether there exist any natural conditions that change the tension of the plasma membrane in yeast. One of these conditions may induce a decrease in the levels of sphingolipids that constitute the yeast plasma membrane together with glycerophospholipids and ergosterols. The initial step in the biosynthetic pathway of sphingolipids is catalyzed by serine palmitoyltransferase. The activity of this enzyme is negatively regulated by Orm1 and its paralog Orm2, the functions of which are controlled through the phosphorylation by Ypk1, a TORC2 substrate, at Ser^{51}, Ser^{52}, and Ser^{53} in Orm1, and Ser^{46}, Ser^{47}, and Ser^{48} in Orm2 [93, 94]. Myriocin is a potent inhibitor of serine palmitoyltransferase; therefore, the treatment of yeast cells with this chemical reduces the production of sphingolipids, which causes feedback regulation to activate sphingolipid biosynthesis through

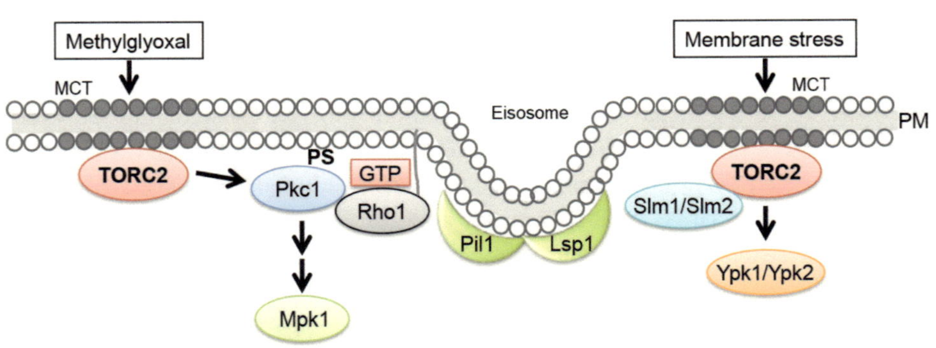

Figure 3. Activation of TORC2 signaling in *S. cerevisiae*.

the stimulation of TORC2-Ypk1 signaling. Orm1/2 is subsequently phosphorylated, and its inhibitory effects on serine palmitoyltransferase are then compromised. Aureobasidin A, a cyclic depsipeptide antibiotic drug, exerts similar effects on the yeast plasma membrane in terms of altering membrane tension because this chemical inhibits the synthesis of inositol-phosphoceramide, one of the sphingolipid species in yeast. Aureobasidin A and myriocin consistently induce the phosphorylation of Ypk1 at Thr^{662}, a target site of TORC2 [94].

4.3.2. *Activation of TORC2 signaling by the metabolic cue methylglyoxal*

In contrast to mammals, which possess many isozymes of protein kinase C and its related kinases, Pkc1 is the sole protein kinase C in budding yeast. Pkc1 is involved in numerous pivotal biological functions including the organization of the actin cytoskeleton and the maintenance of cell wall integrity (CWI). The Mpk1 MAPK cascade lies downstream of Pkc1, and the Pkc1-Mpk1 MAPK cascade constitutes the main stream of the CWI pathway [95]. Chemicals that provoke cell wall damage such as Congo red or heat-shock stress activate the CWI pathway. The small GTPase Rho1 plays a crucial role in the heat-shock stress-induced activation of the CWI pathway, that is, the transmembrane proteins Wsc1 and Mid2 on the plasma membrane sense heat shock and interact with Rom2, a GEF toward Rho1, to load GTP to Rho1. $Rho1^{GTP}$ physically interacts with Pkc1 to communicate the signal to the downstream Mpk1 MAPK cascade [96, 97]. A recent study reported that phosphatidylserine, one of the major glycerophospholipids prevailing in the plasma membrane, mediates the physical interaction between Pkc1 and $Rho1^{GTP}$ [98, 99]. On the other hand, methylglyoxal, a typical 2-oxoaldehyde derived from glycolysis [100], also activates the Pkc1-Mpk1 MAPK cascade; however, the methylglyoxal-induced activation of this pathway is not dependent on Wsc1/Mid2, whereas Rho1 is indispensable [101]. Besides Ypk1 and Ypk2, Pkc1 has also been identified as a direct substrate of TORC2 in *S. cerevisiae*, that is, Thr^{1125} within the turn motif and Ser^{1143} within the hydrophobic motif in Pkc1 are phosphorylated by TORC2 [101]. Methylglyoxal enhanced the phosphorylation levels of Pkc1 at Ser^{1143} in a TORC2-dependent manner [101] (**Figure 3**).

The methylglyoxal-induced activation of TORC2 is conserved in mammalian cells, that is, the phosphorylation levels of Ser^{473} within the hydrophobic motif in Akt, a substrate of mTORC2, were enhanced following the treatment of mouse 3 T3-L1 cells with methylglyoxal [101]. Collectively, these findings demonstrate that methylglyoxal activates (m)TORC2 signaling in yeast and mammalian cells; however, the underlying mechanisms have not yet been elucidated. Since methylglyoxal is a naturally occurring ubiquitous metabolite and is involved in type 2 diabetes and its complications [100], its involvement in the activation of (m)TORC2 signaling is of considerable interest in order to obtain insights into not only novel activation mechanisms of TORC2 but also the physiological significance of methylglyoxal.

4.3.3. *Activation of mTORC2 signaling by growth factor*

In mammalian cells, physiological cues for the activation of mTORC2 signaling are insulin and insulin-like growth factors [102]. Upon the capture of ligands by tyrosine kinase-type receptors, tyrosine-phosphorylated IRS (*i*nsulin *r*eceptor *s*ubstrate) undergoes the activation of phosphatidylinositol 3-kinase, which enhances the levels of phosphatidylinositol

(3,4,5)-trisphosphate (PtdIns(3,4,5)P_3). Two events are subsequently induced by this phosphoinositide near the plasma membrane: that is, the activation of PDK (***p***hosphoinositide-***d***ependent ***k***inase) and the recruitment of Akt to the plasma membrane in which the PH domain of Akt binds to PtdIns(3,4,5)P_3. In turn, Akt at Thr^{308} within the activation loop and Ser^{473} within the hydrophobic motif are phosphorylated by PDK and mTORC2, respectively; however, the mechanisms by which insulin activates mTORC2 remain obscure.

5. Concluding remarks

Laboratory conditions for culturing yeast may be adequate for yeast cells to maintain cellular activities because ample amounts of glucose and amino acids are typically supplied in media. By contrast, nutritional conditions surrounding yeast cells that exist in the natural world are harsh and variable. Yeast cells have evolved mechanisms for sensing changes in nutritional conditions and transitioning the metabolic status and gene expression profile to adapt efficiently and survive inhospitable conditions. The TOR signaling system had been acquired as one of these signal network systems and has been evolutionarily conserved among eukaryotes. In higher eukaryotes, such as humans, dysfunctions in the TOR signaling network closely correlate with pathological conditions including diabetes, cancer, obesity, and neurodegeneration [1]; therefore, TOR is a target from a clinical point of view. Upstream and downstream processes of TORC1 signaling have been extensively investigated because rapamycin, a potent inhibitor for TORC1, was available. By contrast, studies on TORC2 signaling appear to be challenging because of the absence of TORC2-specific inhibitors. However, yeast was always a vanguard from the beginning of TOR studies (TOR was discovered by genetic screening using yeast) and will continue to be so in the future. Many issues remain to be solved in TOR signaling; however, since TOR is a central player in cell growth, studies on TOR will be nothing less than a study of the living system itself. Investigations on TOR will provide many insights for understanding "life."

Author details

Yoshiharu Inoue* and Wataru Nomura

*Address all correspondence to: y_inoue@kais.kyoto-u.ac.jp

Graduate School of Agriculture, Kyoto University, Uji, Kyoto, Japan

References

[1] Laplante M, Sabatini DM. mTOR signaling in growth control and disease. Cell. 2012;**149**:274-293. DOI: 10.1016/j.cell.2012.03.017

[2] Gaubitz C, Prouteau M, Kusmider B, Loewith R. TORC2 structure and function. Trends in Biochemical Sciences. 2016;**41**:532-545. DOI: 10.1016/j.tibs.2016.04.001

[3] Eltschinger S, Loewith R. TOR complexes and the maintenance of cellular homeostasis. Trends in Cell Biology. 2016;**26**:148-159. DOI: 10.1016/j.tcb.2015.10.003

[4] González A, Hall MN. Nutrient sensing and TOR signaling in yeast and mammals. The EMBO Journal. 2017;**36**:397-408. DOI: 10.15252/embj.201696010

[5] Martel RR, Klicius J, Galet S. Inhibition of the immune response by rapamycin, a new antifungal antibiotic. Canadian Journal of Physiology and Pharmacology. 1977;**55**:48-51. DOI: 10.1139/y77-007

[6] Guba M, von Breitenbuch P, Steinbauer M, Koehl G, Flege S, Hornung M, Bruns CJ, Zuelke C, Farkas S, Anthuber M, Jauch KW, Geissler EK. Rapamycin inhibits primary and metastatic tumor growth by antiangiogenesis: Involvement of vascular endothelial growth factor. Nature Medicine. 2002;**8**:128-135. DOI: 10.1038/nm0202-128

[7] Schreiber SL. Chemistry and biology of the immunophilins and their immunosuppressive ligands. Science. 1991;**251**:283-287. DOI: 10.1126/science.1702904

[8] Heitman J, Movva NR, Hiestand PC, Hall MN. FK506-binding protein proline rotamase is a target for the immunosuppressive agent FK506 in *Saccharomyces cerevisiae*. Proceedings of the National Academy of Sciences of the United States of America. 1991;**88**:1948-1952

[9] Koltin Y, Faucette L, Bergsma DJ, Levy MA, Cafferkey R, Koser PL, Johnson RK, Livi GP. Rapamycin sensitivity in *Saccharomyces cerevisiae* is mediated by a peptidyl-prolyl *cis-trans* isomerase related to human FK506-binding protein. Molecular and Cellular Biology. 1991;**11**:1718-1723. DOI: 10.1128/MCB.11.3.1718

[10] Tanida I, Yanagida M, Maki N, Yagi S, Namiyama F, Kobayashi T, Hayano T, Takahashi N, Suzuki M. Yeast cyclophilin-related gene encodes a nonessential second peptidyl-prolyl *cis-trans* isomerase associated with the secretory pathway. Transplantation Proceedings. 1991;**3**:2856-2861

[11] Wiederrecht G, Brizuela L, Elliston K, Sigal NH, Siekierka JJ. *FKB1* encodes a nonessential FK 506-binding protein in *Saccharomyces cerevisiae* and contains regions suggesting homology to the cyclophilins. Proceedings of the National Academy of Sciences of the United States of America. 1991;**88**:1029-1033. DOI: 10.1073/pnas.88.3.1029

[12] Heitman J, Movva NR, Hall MN. Targets for cell cycle arrest by the immunosuppressant rapamycin in yeast. Science. 1991;**253**:905-909. DOI: 10.1126/science.1715094

[13] Cafferkey R, Young PR, McLaughlin MM, Bergsma DJ, Koltin Y, Sathe GM, Faucette L, Eng WK, Johnson RK, Livi GP. Dominant missense mutations in a novel yeast protein related to mammalian phosphatidylinositol 3-kinase and VPS34 abrogate rapamycin cytotoxicity. Molecular and Cellular Biology. 1993;**13**:6012-6023. DOI: 10.1128/MCB.13.10.6012

[14] Kunz J, Henriquez R, Schneider U, Deuter-Reinhard M, Movva NR, Hall MN. Target of rapamycin in yeast, TOR2, is an essential phosphatidylinositol kinase homolog required for G1 progression. Cell. 1993;**73**:585-596. DOI: 10.1016/0092-8674(93)90144-F

[15] Stan R, McLaughlin MM, Cafferkey R, Johnson RK, Rosenberg M, Livi GP. Interaction between FKBP12-rapamycin and TOR involves a conserved serine residue. The Journal of Biological Chemistry. 1994;**269**:32027-32030

[16] Lorenz MC, Heitman J. TOR mutations confer rapamycin resistance by preventing interaction with FKBP12-rapamycin. The Journal of Biological Chemistry. 1995;**270**:27531-27537. DOI: 10.1074/jbc.270.46.27531

[17] Zheng XF, Florentino D, Chen J, Crabtree GR, Schreiber SL. TOR kinase domains are required for two distinct functions, only one of which is inhibited by rapamycin. Cell. 1995;**82**:121-130. DOI: 10.1016/0092-8674(95)90058-6

[18] Schmelzle T, Hall MN. TOR, a central controller of cell growth. Cell. 2000;**103**:253-262. DOI: 10.1016/S0092-8674(00)00117-3

[19] Alarcon CM, Heitman J, Cardenas ME. Protein kinase activity and identification of a toxic effector domain of the target of rapamycin TOR proteins in yeast. Molecular Biology of the Cell. 1999;**10**:2531-2546. DOI: 10.1091/mbc.10.8.2531

[20] Bosotti R, Isacchi A, Sonnhammer EL. FAT: A novel domain in PIK-related kinases. Trends in Biochemical Sciences. 2000;**25**:225-227. DOI: 10.1016/S0968-0004(00)01563-2

[21] Dames SA, Mulet JM, Rathgeb-Szabo K, Hall MN, Grzesiek S. The solution structure of the FATC domain of the protein kinase target of rapamycin suggests a role for redox-dependent structural and cellular stability. The Journal of Biological Chemistry. 2005;**280**:20558-20564. DOI: 10.1074/jbc.M501116200

[22] Helliwell SB, Wagner P, Kunz J, Deuter-Reinhard M, Henriquez R, Hall MN. TOR1 and TOR2 are structurally and functionally similar but not identical phosphatidylinositol kinase homologues in yeast. Molecular Biology of the Cell. 1994;**5**:105-118. DOI: 10.1091/mbc.5.1.105

[23] Rohde J, Heitman J, Cardenas ME. The TOR kinases link nutrient sensing to cell growth. The Journal of Biological Chemistry. 2001;**276**:9583-9586. DOI: 10.1074/jbc.R000034200

[24] Barbet NC, Schneider U, Helliwell SB, Stansfield I, Tuite MF, Hall MN. TOR controls translation initiation and early G1 progression in yeast. Molecular Biology of the Cell. 1996;**7**:25-42. DOI: 10.1091/mbc.7.1.25

[25] Gaubitz C, Oliveira TM, Prouteau M, Leitner A, Karuppasamy M, Konstantinidou G, Rispal D, Eltschinger S, Robinson GC, Thore S, Aebersold R, Schaffitzel C, Loewith R. Molecular basis of the rapamycin insensitivity of target of rapamycin complex 2. Molecular Cell. 2015;**58**:977-988. DOI: 10.1016/j.molcel.2015.04.031

[26] Loewith R, Jacinto E, Wullschleger S, Lorberg A, Crespo JL, Bonenfant D, Oppliger W, Jenoe P, Hall MN. Two TOR complexes, only one of which is rapamycin sensitive, have distinct roles in cell growth control. Molecular Cell. 2002;**10**:457-468. DOI: 10.1016/S1097-2765(02)00636-6

[27] Wedaman KP, Reinke A, Anderson S, Yates J 3rd, McCaffery JM, Powers T. Tor kinases are in distinct membrane-associated protein complexes in *Saccharomyces cerevisiae*. Molecular Biology of the Cell 2003;**14**:1204-1220. DOI: 10.1091/mbc.E02-09-0609

[28] Reinke A, Anderson S, McCaffery JM, Yates J 3rd, Aronova S, Chu S, Fairclough S, Iverson C, Wedaman KP, Powers T. Tor complex 1 includes a novel component, Tco89p (YPL180w), and cooperates with Ssd1p to maintain cellular integrity in *Saccharomyces cerevisiae*. The Journal of Biological Chemistry 2004;**279**:14752-14762. DOI: 10.1074/jbc.M313062200

[29] Yip CK, Murata K, Walz T, Sabatini DM, Kang SA. Structure of the human mTOR complex I and its implications for rapamycin inhibition. Molecular Cell. 2010;**38**:768-774. DOI: 10.1016/j.molcel.2010.05.017

[30] Dubouloz F, Deloche O, Wanke V, Cameroni E, De Virgilio C. The TOR and EGO protein complexes orchestrate microautophagy in yeast. Molecular Cell. 2005;**19**:15-26. DOI: 10.1016/j.molcel.2005.05.020

[31] Gao M, Kaiser CA. A conserved GTPase-containing complex is required for intracellular sorting of the general amino-acid permease in yeast. Nature Cell Biology. 2006;**8**:657-667. DOI: 10.1038/ncb1419

[32] Gong R, Li L, Liu Y, Wang P, Yang H, Wang L, Cheng J, Guan KL, Xu Y. Crystal structure of the Gtr1p-Gtr2p complex reveals new insights into the amino acid-induced TORC1 activation. Genes & Development. 2011;**25**:1668-1673. DOI: 10.1101/gad.16968011

[33] Sekiguchi T, Hirose E, Nakashima N, Ii M, Nishimoto T. Novel G proteins, Rag C and Rag D, interact with GTP-binding proteins, Rag A and Rag B. The Journal of Biological Chemistry. 2001;**276**:7246-7257. DOI: 10.1074/jbc.M004389200

[34] Nakashima N, Noguchi E, Nishimoto T. *Saccharomyces cerevisiae* putative G protein, Gtr1p, which forms complexes with itself and a novel protein designated as Gtr2p, negatively regulates the Ran/Gsp1p G protein cycle through Gtr2p. Genetics. 1999;**152**:853-867

[35] Powis K, Zhang T, Panchaud N, Wang R, De Virgilio C, Ding J. Crystal structure of the Ego1-Ego2-Ego3 complex and its role in promoting Rag GTPase-dependent TORC1 signaling. Cell Research. 2015;**25**:1043-1059. DOI: 10.1038/cr.2015.86

[36] Binda M, Péli-Gulli MP, Bonfils G, Panchaud N, Urban J, Sturgill TW, Loewith R, De Virgilio C. The Vam6 GEF controls TORC1 by activating the EGO complex. Molecular Cell. 2009;**35**:563-573. DOI: 10.1016/j.molcel.2009.06.033

[37] Ashrafi K, Farazi TA, Gordon JI. A role for *Saccharomyces cerevisiae* fatty acid activation protein 4 in regulating protein *N*-myristoylation during entry into stationary phase. The Journal of Biological Chemistry. 1998;**273**:25864-25874. DOI: 10.1074/jbc.273.40.25864

[38] Nadolski MJ, Linder ME. Molecular recognition of the palmitoylation substrate Vac8 by its palmitoyltransferase Pfa3. The Journal of Biological Chemistry. 2009;**284**:17720-17730. DOI: 10.1074/jbc.M109.005447

[39] Roth AF, Wan J, Bailey AO, Sun B, Kuchar JA, Green WN, Phinney BS, Yates JR 3rd, Davis NG. Global analysis of protein palmitoylation in yeast. Cell 2006;**125**:1003-1013. DOI: 10.1016/j.cell.2006.03.042

[40] Nada S, Hondo A, Kasai A, Koike M, Saito K, Uchiyama Y, Okada M. The novel lipid raft adaptor p18 controls endosome dynamics by anchoring the MEK-ERK pathway to late endosomes. The EMBO Journal. 2009;**28**:477-489. DOI: 10.1038/emboj.2008.308

[41] Zhang T, Péli-Gulli MP, Yang H, De Virgilio C, Ding J. Ego3 functions as a homodimer to mediate the interaction between Gtr1-Gtr2 and Ego1 in the EGO complex to activate TORC1. Structure. 2012;**20**:2151-2160. DOI: 10.1016/j.str.2012.09.019

[42] Levine TP, Daniels RD, Wong LH, Gatta AT, Gerondopoulos A, Barr FA. Discovery of new Longin and Roadblock domains that form platforms for small GTPases in Ragulator and TRAPP-II. Small GTPases. 2013;**4**:62-69. DOI: 10.4161/sgtp.24262

[43] Petit CS, Roczniak-Ferguson A, Ferguson SM. Recruitment of folliculin to lysosomes supports the amino acid-dependent activation of Rag GTPases. The Journal of Cell Biology. 2013;**202**:1107-1122. DOI: 10.1083/jcb.201307084

[44] Tsun ZY, Bar-Peled L, Chantranupong L, Zoncu R, Wang T, Kim C, Spooner E, Sabatini DM. The folliculin tumor suppressor is a GAP for the RagC/D GTPases that signal amino acid levels to mTORC1. Molecular Cell. 2013;**52**:495-505. DOI: 10.1016/j.molcel.2013.09.016

[45] Péli-Gulli MP, Sardu A, Panchaud N, Raucci S, De Virgilio C. Amino acids stimulate TORC1 through Lst4-Lst7, a GTPase-activating protein complex for the Rag family GTPase Gtr2. Cell Reports. 2015;**13**:1-7. DOI: 10.1016/j.celrep.2015.08.059

[46] Neklesa TK, Davis RW. A genome-wide screen for regulators of TORC1 in response to amino acid starvation reveals a conserved Npr2/3 complex. PLoS Genetics. 2009;**5**: e1000515. DOI: 10.1371/journal.pgen.1000515

[47] Dokudovskaya S, Rout MP. A novel coatomer-related SEA complex dynamically associates with the vacuole in yeast and is implicated in the response to nitrogen starvation. Autophagy. 2011;**7**:1392-1393. DOI: 10.4161/auto.7.11.17347

[48] Wu X, Tu BP. Selective regulation of autophagy by the Iml1-Npr2-Npr3 complex in the absence of nitrogen starvation. Molecular Biology of the Cell. 2011;**22**:4124-4133. DOI: 10.1091/mbc.E11-06-0525

[49] Panchaud N, Péli-Gulli MP, De Virgilio C. SEACing the GAP that nEGOCiates TORC1 activation: Evolutionary conservation of Rag GTPase regulation. Cell Cycle. 2013;**12**:2948-2952. DOI: 10.4161/cc.26000

[50] Panchaud N, Péli-Gulli MP, De Virgilio C. Amino acid deprivation inhibits TORC1 through a GTPase-activating protein complex for the Rag family GTPase Gtr1. Science Signaling. 2013;**6**:ra42. DOI: 10.1126/scisignal.2004112

[51] Dokudovskaya S, Rout MP. SEA you later alli-GATOR—A dynamic regulator of the TORC1 stress response pathway. Journal of Cell Science. 2015;**128**:2219-2228. DOI: 10.1242/jcs.168922

[52] Wolfson RL, Chantranupong L, Saxton RA, Shen K, Scaria SM, Cantor JR, Sabatini DM. Sestrin2 is a leucine sensor for the mTORC1 pathway. Science. 2016;**351**:43-48. DOI: 10.1126/science.aab2674

[53] Chantranupong L, Scaria SM, Saxton RA, Gygi MP, Shen K, Wyant GA, Wang T, Harper JW, Gygi SP, Sabatini DM. The CASTOR proteins are arginine sensors for the mTORC1 pathway. Cell. 2016;**165**:153-164. DOI: 10.1016/j.cell.2016.02.035

[54] Parmigiani A, Nourbakhsh A, Ding B, Wang W, Kim YC, Akopiants K, Guan KL, Karin M, Budanov AV. Sestrins inhibit mTORC1 kinase activation through the GATOR complex. Cell Reports. 2014;**9**:1281-1291. DOI: 10.1016/j.celrep.2014.10.019

[55] Chantranupong L, Wolfson RL, Orozco JM, Saxton RA, Scaria SM, Bar-Peled L, Spooner E, Isasa M, Gygi SP, Sabatini DM. The Sestrins interact with GATOR2 to negatively regulate the amino-acid-sensing pathway upstream of mTORC1. Cell Reports. 2014;**9**:1-8. DOI: 10.1016/j.celrep.2014.09.014

[56] Kim JS, Ro SH, Kim M, Park HW, Semple IA, Park H, Cho US, Wang W, Guan KL, Karin M, Lee JH. Sestrin2 inhibits mTORC1 through modulation of GATOR complexes. Scientific Reports. 2015;**5**:9502. DOI: 10.1038/srep09502

[57] Kamada Y. Novel tRNA function in amino acid sensing of yeast Tor complex1. Genes to Cells. 2017;**22**:135-147. DOI: 10.1111/gtc.12462

[58] Stracka D, Jozefczuk S, Rudroff F, Sauer U, Hall MN. Nitrogen source activates TOR (target of rapamycin) complex 1 via glutamine and independently of Gtr/Rag proteins. The Journal of Biological Chemistry. 2014;**289**:25010-25020. DOI: 10.1074/jbc.M114.574335

[59] Thomas JD, Zhang YJ, Wei YH, Cho JH, Morris LE, Wang HY, Zheng XFS. Rab1A is an mTORC1 activator and a colorectal oncogene. Cancer Cell. 2014;**6**:754-769. DOI: 10.1016/j.ccell.2016.06.014

[60] Jewell JL, Kim YC, Russell RC, Yu FX, Park HW, Plouffe SW, Tagliabracci VS, Metabolism GKL. Differential regulation of mTORC1 by leucine and glutamine. Science. 2015;**347**:194-198. DOI: 10.1126/science.1259472

[61] Tanigawa M, Maeda T. An *in vitro* TORC1 kinase assay that recapitulates the Gtr-independent glutamine-responsive TORC1 activation mechanism on yeast vacuoles. Molecular and Cellular Biology. 2017;**37**:e00075-e00017. DOI: 10.1128/MCB.00075-17

[62] Menon S, Dibble CC, Talbott G, Hoxhaj G, Valvezan AJ, Takahashi H, Cantley LC, Manning BD. Spatial control of the TSC complex integrates insulin and nutrient regulation of mTORC1 at the lysosome. Cell. 2014;**156**:771-785. DOI: 10.1016/j.cell.2013.11.049

[63] Demetriades C, Doumpas N, Teleman AA. Regulation of TORC1 in response to amino acid starvation via lysosomal recruitment of TSC2. Cell. 2014;**156**:786-799. DOI: 10.1016/j.cell.2014.01.024

[64] Carroll B, Maetzel D, Maddocks OD, Otten G, Ratcliff M, Smith GR, Dunlop EA, Passos JF, Davies OR, Jaenisch R, Tee AR, Sarkar S, Korolchuk VI. Control of TSC2-Rheb signaling axis by arginine regulates mTORC1 activity. eLife. 2016;**5**:e11058. DOI: 10.7554/eLife.11058

[65] Urano J, Comiso MJ, Guo L, Aspuria PJ, Deniskin R, Tabancay AP Jr, Kato-Stankiewicz J, Tamanoi F. Identification of novel single amino acid changes that result in hyperactivation of the unique GTPase, Rheb, in fission yeast. Molecular Microbiology 2005;**58**:1074-1086. DOI: 10.1111/j.1365-2958.2005.04877.x

[66] Nakase Y, Fukuda K, Chikashige Y, Tsutsumi C, Morita D, Kawamoto S, Ohnuki M, Hiraoka Y, Matsumoto T. A defect in protein farnesylation suppresses a loss of *Schizosaccharomyces pombe tsc2*$^{+}$, a homolog of the human gene predisposing to tuberous sclerosis complex. Genetics. 2006;**173**:569-578. DOI: 10.1534/genetics.106.056895

[67] van Slegtenhorst M, Mustafa A, Henske EP. Pas1, a G1 cyclin, regulates amino acid uptake and rescues a delay in G1 arrest in Tsc1 and Tsc2 mutants in *Schizosaccharomyces pombe*. Human Molecular Genetics. 2005;**14**:2851-2858. DOI: 10.1093/hmg/ddi317

[68] Uritani M, Hidaka H, Hotta Y, Ueno M, Ushimaru T, Toda T. Fission yeast Tor2 links nitrogen signals to cell proliferation and acts downstream of the Rheb GTPase. Genes to Cells. 2006;**11**:1367-1379. DOI: 10.1111/j.1365-2443.2006.01025.x

[69] Mach KE, Furge KA, Albright CF. Loss of Rhb1, a Rheb-related GTPase in fission yeast, causes growth arrest with a terminal phenotype similar to that caused by nitrogen starvation. Genetics. 2000;**155**:611-622

[70] Yang W, Tabancay AP Jr, Urano J, Tamanoi F. Failure to farnesylate Rheb protein contributes to the enrichment of G0/G1 phase cells in the *Schizosaccharomyces pombe* farnesyltransferase mutant. Molecular Microbiology 2001;**41**:1339–1347. DOI: 10.1046/j.1365-2958.2001.02599.x

[71] Urano J, Sato T, Matsuo T, Otsubo Y, Yamamoto M, Tamanoi F. Point mutations in TOR confer Rheb-independent growth in fission yeast and nutrient-independent mammalian TOR signaling in mammalian cells. Proceedings of the National Academy of Sciences of the United States of America. 2007;**104**:3514-3519. DOI: 10.1073/pnas.0608510104

[72] Berchtold D, Walther TC. TORC2 plasma membrane localization is essential for cell viability and restricted to a distinct domain. Molecular Biology of the Cell. 2009;**20**:1565-1575. DOI: 10.1091/mbc.E08-10-1001

[73] Cameron AJ, Linch MD, Saurin AT, Escribano C, Parker PJ. mTORC2 targets AGC kinases through Sin1-dependent recruitment. The Biochemical Journal. 2011;**439**:287-297. DOI: 10.1042/BJ20110678

[74] Liao HC, Chen MY. Target of rapamycin complex 2 signals to downstream effector yeast protein kinase 2 (Ypk2) through adheres-voraciously-to-target-of-rapamycin-2 protein 1 (Avo1) in *Saccharomyces cerevisiae*. The Journal of Biological Chemistry. 2012;**287**:6089-6099. DOI: 10.1074/jbc.M111.303701

[75] Wullschleger S, Loewith R, Oppliger W, Hall MN. Molecular organization of target of rapamycin complex 2. The Journal of Biological Chemistry. 2005;**280**:30697-30704. DOI: 10.1074/jbc.M505553200

[76] Tewari R, Bailes E, Bunting KA, Coates JC. Armadillo-repeat protein functions: Questions for little creatures. Trends in Cell Biology. 2010;**20**:470-481. DOI: 10.1016/j.tcb.2010.05.003

[77] Gatherar IM, Pollerman S, Dunn-Coleman N, Turner G. Identification of a novel gene *hbrB* required for polarised growth in *Aspergillus nidulans*. Fungal Genetics and Biology. 2004;**41**:463-471. DOI: 10.1016/j.fgb.2003.12.004

[78] Berchtold D, Piccolis M, Chiaruttini N, Riezman I, Riezman H, Roux A, Walther TC, Loewith R. Plasma membrane stress induces relocalization of Slm proteins and activation of TORC2 to promote sphingolipid synthesis. Nature Cell Biology. 2012;**14**:542-547. DOI: 10.1038/ncb2480

[79] Colicelli J, Nicolette C, Birchmeier C, Rodgers L, Riggs M, Wigler M. Expression of three mammalian cDNAs that interfere with RAS function in *Saccharomyces cerevisiae*. Proceedings of the National Academy of Sciences of the United States of America. 1991;**88**:2913-2917. DOI: 10.1073/pnas.88.7.2913

[80] Sarbassov DD, Ali SM, Sengupta S, Sheen JH, Hsu PP, Bagley AF, Markhard AL, Sabatini DM. Prolonged rapamycin treatment inhibits mTORC2 assembly and Akt/PKB. Molecular Cell. 2006;**22**:159-168. DOI: 10.1016/j.molcel.2006.03.029

[81] Tatebe H, Morigasaki S, Murayama S, Zeng CT, Shiozaki K. Rab-family GTPase regulates TOR complex 2 signaling in fission yeast. Current Biology. 2010;**20**:1975-1982. DOI: 10.1016/j.cub.2010.10.026

[82] Saci A, Cantley LC, Carpenter CL. Rac1 regulates the activity of mTORC1 and mTORC2 and controls cellular size. Molecular Cell. 2011;**42**:50-61. DOI: 10.1016/j.molcel.2011.03.017

[83] Cai W, Andres DA. mTORC2 is required for Rit-mediated oxidative stress resistance. PLoS One. 2014;**9**:e115602. DOI: 10.1371/journal.pone.0115602

[84] Niles BJ, Powers T. Tor complex 2–Ypk1 signaling regulates actin polarization via reactive oxygen species. Molecular Biology of the Cell. 2014;**25**:3962-3972. DOI: 10.1091/mbc.E14-06-1122

[85] Hálová L, Du W, Kirkham S, Smith DL, Petersen J. Phosphorylation of the TOR ATP binding domain by AGC kinase constitutes a novel mode of TOR inhibition. The Journal of Cell Biology. 2013;**203**:595-604. DOI: 10.1083/jcb.201305103

[86] Dibble CC, Asara JM, Manning BD. Characterization of Rictor phosphorylation sites reveals direct regulation of mTOR complex 2 by S6K1. Molecular and Cellular Biology. 2009;**29**:5657-5670. DOI: 10.1128/MCB.00735-09

[87] Chen CH, Kiyan V, Zhylkibayev AA, Kazyken D, Bulgakova O, Page KE, Bersimbaev RI, Spooner E, Sarbassov DD. Autoregulation of the mechanistic target of rapamycin (mTOR) complex 2 integrity is controlled by an ATP-dependent mechanism. The Journal of Biological Chemistry. 2013;**288**:27019-27030. DOI: 10.1074/jbc.M113.498055

[88] Liu P, Guo J, Gan W, Wei W. Dual phosphorylation of Sin1 at T86 and T398 negatively regulates mTORC2 complex integrity and activity. Protein & Cell. 2014;**5**:171-177. DOI: 10.1007/s13238-014-0021-8

[89] Swaney DL, Beltrao P, Starita L, Guo A, Rush J, Fields S, Krogan NJ, Villén J. Global analysis of phosphorylation and ubiquitylation cross-talk in protein degradation. Nature Methods. 2013;**10**:676-682. DOI: 10.1038/nmeth.2519

[90] Glidden EJ, Gray LG, Vemuru S, Li D, Harris TE, Mayo MW. Multiple site acetylation of Rictor stimulates mammalian target of rapamycin complex 2 (mTORC2)-dependent phosphorylation of Akt protein. The Journal of Biological Chemistry. 2012;**287**:581-588. DOI: 10.1074/jbc.M111.304337

[91] Masui K, Tanaka K, Akhavan D, Babic I, Gini B, Matsutani T, Iwanami A, Liu F, Villa GR, Gu Y, Campos C, Zhu S, Yang H, Yong WH, Cloughesy TF, Mellinghoff IK, Cavenee WK, Shaw RJ, Mischel PS. mTOR complex 2 controls glycolytic metabolism in glioblastoma through FoxO acetylation and upregulation of c-Myc. Cell Metabolism. 2013;**18**:726-739. DOI: 10.1016/j.cmet.2013.09.013

[92] Douglas LM, Konopka JB. Fungal membrane organization: The eisosome concept. Annual Review of Microbiology. 2014;**68**:377-393. DOI: 10.1146/annurev-micro-091313-103507

[93] Consonni SV, Maurice MM, Bos JL. DEP domains: Structurally similar but functionally different. Nature Reviews. Molecular Cell Biology. 2014;**15**:357-362. DOI: 10.1038/nrm3791

[94] Muir A, Ramachandran S, Roelants FM, Timmons G, Thorner J. TORC2-dependent protein kinase Ypk1 phosphorylates ceramide synthase to stimulate synthesis of complex sphingolipids. eLife. 2014;**3**:e03779. DOI: 10.7554/eLife.03779

[95] Levin DE. Cell wall integrity signaling in *Saccharomyces cerevisiae*. Microbiology and Molecular Biology Reviews. 2005;**69**:262-291. DOI: 10.1128/MMBR.69.2.262-291.2005

[96] Nonaka H, Tanaka K, Hirano H, Fujiwara T, Kohno H, Umikawa M, Mino A, Takai Y. A downstream target of RHO1 small GTP-binding protein is PKC1, a homolog of protein kinase C, which leads to activation of the MAP kinase cascade in *Saccharomyces cerevisiae*. The EMBO Journal. 1995;**14**:5931-5938

[97] Schmitz HP, Lorberg A, Heinisch JJ. Regulation of yeast protein kinase C activity by interaction with the small GTPase Rho1p through its amino-terminal HR1 domain. Molecular Microbiology. 2002;**44**:829-840. DOI: 10.1046/j.1365-2958.2002.02925.x

[98] Nomura W, Ito Y, Inoue Y. Role of phosphatidylserine in the activation of Rho1-related Pkc1 signaling in *Saccharomyces cerevisiae*. Cellular Signalling. 2017;**31**:146-153. DOI: 10.1016/j.cellsig.2017.01.002

[99] Nomura W, Inoue Y. Contribution of phosphatidylserine to Rho1- and Pkc1-related repolarization of the actin cytoskeleton under stressed conditions in *Saccharomyces cerevisiae*. Small GTPases. 2017;**14**:1-7. DOI: 10.1080/21541248.2017.1339766

[100] Inoue Y, Maeta K, Nomura W. Glyoxalase system in yeasts: Structure, function, and physiology. Seminars in Cell & Developmental Biology. 2011;**22**:278-284. DOI: 10.1016/j.semcdb.2011.02.002

[101] Nomura W, Inoue Y. Methylglyoxal activates the target of rapamycin complex 2-protein kinase C signaling pathway in *Saccharomyces cerevisiae*. Molecular and Cellular Biology. 2015;**35**:1269-1280. DOI: 10.1128/MCB.01118-14

[102] Altomare DA, Khaled AR. Homeostasis and the importance for a balance between AKT/mTOR activity and intracellular signaling. Current Medicinal Chemistry. 2012;**19**:3748-3762. DOI: 10.2174/092986712801661130

Chapter 5

Secretion Mechanism across Wall

Aline López-Vargas,
Gurusami Kutralam-Muniasamy, Aldo Amaro-Reyes,
Fermín Pérez-Guevara and
Jorge Noel Gracida Rodríguez

Additional information is available at the end of the chapter

http://dx.doi.org/10.5772/intechopen.70783

Abstract

Yeast organisms are widely explored by humans for different biotechnological applications. During their growth, they need to adapt and interact themselves with the environment medium. For this purpose, organisms uptake nutrients and at the same time secrete different molecules include proteins to extracellular medium. This phenomenon requires the use of specialized structures to regulate entry and exit of molecules called transporters. Two transporters, namely Proteins and Vesicles, are specialized in translocating molecules in and out across the wall. The knowledge of these systems is important and served to bring novel applications of yeast. Taking together, this book chapter is divided into two parts: at first, it primarily accounts on few examples of protein (carbohydrates and peroxisome proteins) and vesicle (intracellular and extracellular vesicles) transporters of yeasts. Second, it deals with the recent advances of yeast applications in diverse area of science.

Keywords: vesicles, symporter, induction, repression, transporters

1. Introduction

For decades, *Saccharomyces cerevisiae* has been the model organism of the lower eukaryotes. The available complete genome favors access to perform possible molecular and genetics researches to understand cell activities of this yeast. The composition and concentration of molecules that conform the outer medium dictates the transporters pattern present in the cellular membrane. The role of molecular transport mechanisms is highly recognized and well-studied. Basically, transport systems can be drawn in two ways; the barrier systems- cell membrane and cell wall. (1) Transport of compounds from inside to outside. (2) And outside to inside. There are many compounds involucrated in the phenomenon: carbohydrates, peptides, some proteins (mucin-type glycopeptides and gycoproteins) and ions (**Figure 1**). As organisms

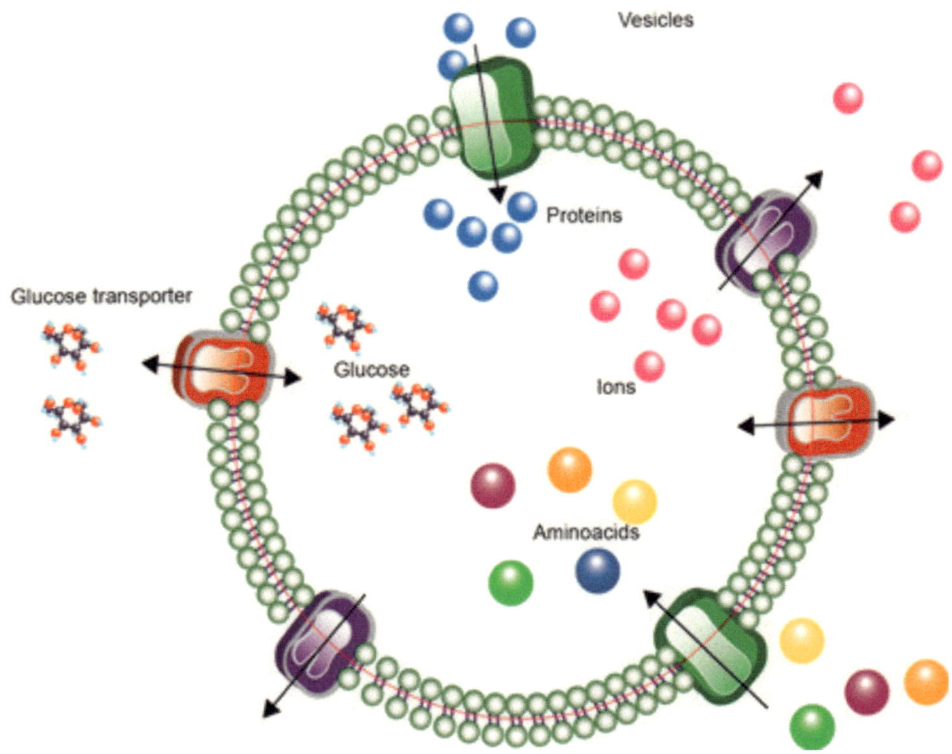

Figure 1. Pictorial representation of diverse transport systems present in the yeast organisms. It includes protein machinery for the transport of biomolecules such as glucose, amino acids, enzymes, ions. Other transporters include vehicles such as vesicles, which are bi-layered liposomes.

like *S. cerevisiae* contains multiple membranous organelles, it is expected to conduct similar types of transport, from cytosol to intracellular organelles across the membrane. Organelle could be any of the following: Nucleus, Mitochondria, Golgi bodies, Endoplasmic reticulum, Peroxisomes. Besides diffusion, various types of transport systems are shown to be involved to either import or export molecules across membranes; we categorize different types reported in the literature. Membrane transporters: Embedded in the membrane are demonstrated to carry out different mechanisms such as a) Protein transport, b) Carbohydrate transport, c) Bilayered membranous vesicles (both intracellular and extracellular vesicles).

2. Hydrocarbons transporters of yeast

There is a great diversity of yeasts and they all require a carbon source to maintain metabolic, physiological and cell growth processes. One of the main nutrients is glucose because it plays a key role regulating the expression of sugar-carrying genes. Yeasts can also consume different types of sugars like xylose, arabinose and under very specific conditions glycerol. These nutrients need to be introduced into the cell whereby yeasts have developed numerous transporters proteins, with similar structures, but with very specific substrate functions and affinities.

2.1. Glucose regulation

Glucose is a substrate of easy metabolism and can act as a signaling molecule depending of its extracellular/intracellular concentration to adjust diverse cellular activities. In *Saccharomyces cerevisiae*, there are two pathways to control glucose consume by regulation of HXT transporter [1].

2.1.1. Pathway of glucose induction Rgt2/Snf3, responsible for its consumption.

Snf3, and Rgt2 are important sugar sensors (not glucose transport) in the *S. cerevisiae*, they are in cell membrane and play key roles, selecting what nutrient to utilize and coordinating expression of sugar transporters. Under low glucose concentration Snf3 sensor elevates the transcription of high-affinity hexose transporter genes, while at high concentrations, Rgt2 sensor promotes low-affinity hexose transporter expression [2]. In both cases, glucose binding to sensors leads their conformation switch and activate a casein kinase I (Yck1/2) which phosphorylates regulators of the glucose-sensing signal transduction pathway Mth1 and Std1 to subsequently be ubiquitylated and degraded [3]. The degradation of Mth1 and Std1 interrupts the interaction between a transcriptional factor Rgt1 and Cyc8/Tup1 to form a general co-repressor complex of expression of HXT genes. Once liberated Rgt1 is phosphorylated therefor HXT genes are expressing [4].

2.1.2. Pathway of glucose repression Snf1/Mth1 negatively regulates genes involved in glucose oxidation and the use of alternative sugars.

Under glucose limitation, there is transcriptional inhibition of hexose transporter genes (HXT) by blocking of their promoter by a repressor complex conformed with Snf1, the complex Cyc8-Tup1 and the Mth1/Std1 [1, 5]. This mechanism is required for the yeast to adapt to glucoses limitation medium.

In *Candida albicans* membrane, the sugar sensor Hht4 (homolog to Snf3 and Rgt2) responses to different levels of sugar by inactivation of a transcriptional repressor Rgt1 that regulates multiple HGT genes encoding hexoses transporters (Hgt2, Hxt10 and Hgt7), this process plays a key role in systemic infections [6]. Hxt4 expression is repressed by high levels of glucose. Snf3p in *Candida glabrata* is essential for growth in low glucose media and plays a role in the induction of several hexoses transporters [7]. Kluyveromyces lactis possesses a system of glucose signaling that also depends of intercellular glucose metabolism, demonstrated in glycolytic mutants whose affection has a direct correlation with the repression of one of its main transporters of glucose Rag1. [8]. **Figure 2** shows the expression pattern of the carbohydrates transporters depending on the glucose concentration.

2.2. Hexoses transporters

The vast majority of yeast carbohydrate transporters belong to the Major Facilitator Superfamily (MFS) and the hexoses transporters (HXT) of *S. cerevisiae* have been extensively studied. HXT family is conformed of 17 putative membrane proteins with a high similarity, but different affinities, [9]. A deletion assay has demonstrated that HXT1–4, HXT6–7 are the mayor functional transporters in glucose or fructose [10]. Some transporters can transport xylose efficiently but there is inhibition by glucose presence because they have a clear preference for this sugar [11].

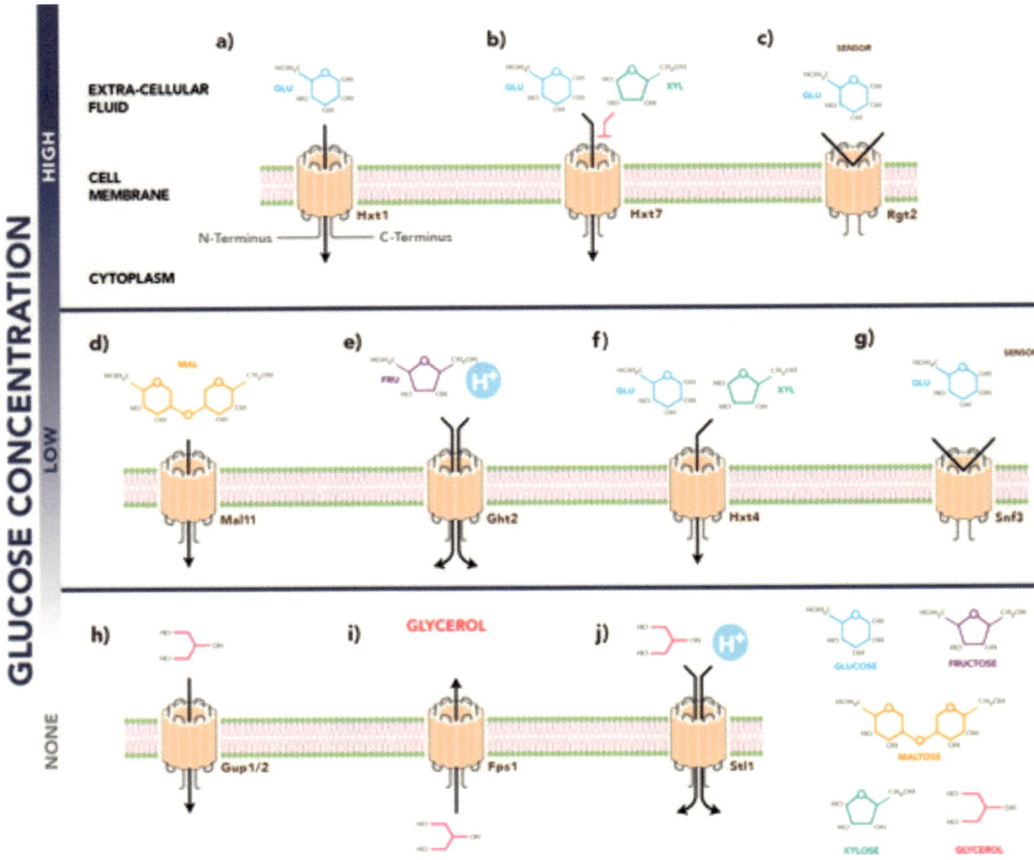

Figure 2. Expression of transporters in yeasts dependent on the concentration of glucose in the medium. At high concentrations of Glu: a) Hxt1 main transporter of Glu, b) Hxt7 has inhibition in xylose transport; c) Rgt2 is a sensor that induces expression of transporters required at high concentrations of glucose. At low concentrations of Glu: d) Mal11 transports α-glusides as maltose, e) Ght2 a symporter proton allows the consumption of fructose, f) decreases inhibition of xylose transport and initiates its consumption, g) Snf3 is a sensor that induces expression of required transporters at low glucose concentrations. In the absence of sugars, yeasts use alternative carbon sources via carriers h) Gup1 or 2 and j) St1. Under osmotic stress, equilibrium is maintained by expelling glycerol via carrier i) Fps1.

All transporters are expressed in several specific conditions. Hxt1 has a low affinity and is expressed in high glucose levels to control carbon flux; in some cases the affinity of the transporters can be modulated to adapt to consumption needs, as Gal2 and Hxt2, they switch affinity to regulate specific transport. Some transporters (Hxt 8–17) are transcribed at low levels and cannot support the demand of nutrients by themselves and in the particular case of Hxt12 does not transport glucose. HXT family has the ability to translocate other sugars such as fructose, which also covers a role of expression regulation [9, 10]. In Schizosaccharomyces pompe, the symporter Ght2 has better affinity for fructose instead of glucose [12].

Specific fructose symporter (Fsy1) has been described to function as a proton symporter; this transporter is able to discriminate between fructose and other hexoses in *Saccharomyces pastorianus* [13]. The symport transport of fructose occurs when cells growth in low fructose (0.5%) medium and exhibits an unusual fructose:H^+ stoichiometry of 1:2 [14].

Frt1, from *Kluyveromyces lactis* encoding a fructose transport protein of high affinity to fructose that acts as a proton-coupled symporter dependent of energy and is rather specific for fructose. This transporter is also induced by galactose, although in a lesser extent. It remains to be determined whether Frt1 transcription is under control of the Rag4 glucose sensor [15].

Yeast *Zygosaccharomyces bailii* uptake fructose mainly at high sugar concentrations by a specific transporter system with high capacity and low affinity and a non-specific transporter with low-capacity and high-affinity that also transports glucose, similar results were also found for *Z. rouxii* [16]. Galactose is another sugar that is uptake by yeast, transporter Gal2 who shows homology with HXT but is not a specific-transporter and presents two conformational state for low and high affinity. In mutant of transporter Lac2 of *K. lactis* lost the capacity of lactose consumption and also for galactose indicating that Lac2 can transport galactose [17].

2.3. Pentoses transporters

Glucose is preferentially transported into the cell due to a 100-fold lower affinity of xylose for the transporters. In *S. cerevisiae* some non-specific Hxt transporters are able to transport xylose only when glucose is absent or in concentrations below 5 g/L. Hxt7 has a low affinity to xylose and presents efficient transport for this sugar [11].

Candida intermedia shows to grow well on xylose, the transport of this sugar is carried on by two different transport systems: a Gxf1 glucose/xylose facilitator 1 with low affinity, it is constitutive expressed and on the other hand a Gxs1 glucose/xylose symporter 1 whit a high affinity to xylose is repressed in the presence of glucose [18]. In [19] was detected a very weak growth in complementation of xylose for YHT1 and YHT6 (genes from *Yarrowia lipolytica*) in an hxt-null mutant of *S. cerevisiae* engineered for use of xylose.

S. cerevisiae lacks of arabinose specific transporters, however Gal2 can be transported at a slow rate, other yeast as *Scheffersomyces stipitis* have an AraT to uptake arabinose and apparently does not facilitate hexoses transport [20].

2.4. α-glucosides transporters

MAL loci contains genes necessary for the transport and consumption of maltose as MALx1 which encodes a maltose permease with low affinity and MALx3 encoding a positive regulatory protein of these genes in the presence of maltose, a clear example would be maltotriose/maltose: symporter Mal61 encoded by MAL61 and a positive regulatory protein encoded by MAL63 [21]. In yeast there are maltose transporters with high and low affinity, for example MAL11, MAL21 and Mal 61 have high affinities to maltose (Michaelis constant (Km): 2–4 mM) and can carry other sugars as turanose but cannot convey maltotriose. Atg1:H^+ is a symporter transporter capable of transporting a wide variety of α-glucosides (trehalose, sucrose, maltose, α-methyl-glucoside, maltotriose) in *S. cerevisiae* [22].

2.5. Glycerol transporters

Polyols like glycerol are used as osmoprotectants by many organisms; yeasts accumulate glycerol under high osmolality conditions. Fps1 glycerol efflux facilitator in *S. cerevisiae* is essential maintaining the balance in hypoosmotic changes, this transmembranal protein contain a cytosolic terminal domains that is important regulating glycerol flux through the channel [23].

Another glycerol:H^+ proton symport transporters like Stl1 are expressed transitorily and activated when all sugar is consumed and the yeast enters into diauxic shift, during this, major changes in gene expression alter the fermentative to oxidative metabolism, allowing to utilize the produced ethanol and glycerol before entry into the stationary phase. Stl1 was inactive in the presence of glucose [24]. By homology analysis with Stl1 from *S. cerevisiae*, the Gt1 *of Scheffersomyces stipitis* demonstrated to be a glycerol transporter that is active when the medium contains ethanol and absence of sugars [25].

2.6. Inositol transporter

Inositol transporters ITR1 and ITR2 (from *S. cerevisiae*) are located in the plasma membrane and accept myo-inositol, both have similar affinities (ITR1: Km = 100 μM, ITR2: Km = 140 μM). However, there is insufficient inositol uptake when only ITR2 is present and there for ITR1 appears to be responsible for inositol uptake because ITR1 is highly transcripted. The ITR2 transporter of Schizosaccharomyces pombe (inositol auxotroph) is essential for regular cell growing, this transporter contains 12 intermembranial domains whit two sugar-transport motifs typical for HXT and shows similarity whit *S. cerevisiae* inositol transporters. The mRNA levels of itr2 gene are also repressed by glucose [26].

All transporters mentioned have the transport of carbohydrates in common, but they present variation on substrate affinity that can be classified in low (Km: >40 mM) and high (Km <40 mM) affinity, this feature leads to control carbon flux; therefore at high substrate concentrations the expression of low affinity transporters is induced. One way to measure carbohydrate transport rates (uptake) is by scintillation assay, where studied strains that express the transporter of interest. It is harvested and transferred to a substrate-free buffer to subsequently expose them to a solution of known concentration of the radioactively labeled carbohydrate of interest for a defined period of time, then, the cells are filtered and washed with the same buffer, after, the remanent is analized by a liquid scintillation counter. The difference between the radioactivity data of the initial substrate and the remaining concentrations, allows substrate consumption quantification per unit time; this information can also be integrated into an enzymatic modeling or nonlinear regression analysis to obtain kinetic parameters of Km and maximal initial uptake speed (Vmax). **Table 1** presents a list of diverse characterized yeast and transporters.

3. Protein transport

Membranous and non-membranous proteins are the indispensable machinery for the cells life. Membranous proteins are integral and peripheral membrane proteins that include transporters (sugars, ions), GTP binding proteins, cell wall synthesizing proteins. While, non-membranous proteins are metabolic proteins, transcription factors and so on. Most proteins are usually encoded in the nucleus and synthesized in the free ribosomes of cytoplasm [39].

http://dx.doi.org/10.5772/intechopen.70783

Species	Transporter	Hexoses	Km	Pentoses	Km	Other	Km	Regulation	Reference
Saccharomyces cerevisiae									
	Hxt1	Glu > Fru	Low	Xyl	Low			Induced by high glucose level Does not transport xylose as unique carbon source	[9]*
	Hxt2	Glu	Mod					Induced by low glucose levels	
								Repressed by high glucose levels	
	Hxt3	Glu > Fru	Low					Induced by high glucose levels	
	Hxt4	Glu > Fru	Low	Xyl	Low	As(OH)3	NR	Induced by low glucose levels	
								Repressed by high glucose levels	
	Hxt5	Glu > Fru	Low	Xyl	Low			Regulated by cellular growth	
	Hxt6	Glu > Fru > Man	High	Xyl		Maltose	NR	Induced slightly at low glucose concentrations Highly induced in non-fermentable substrates	
	Hxt7	Glu	High	Xyl	Low			Repression by high glucose levels. It varies only in 2 amino acids with Hxt6	[11]
	Atg1					Treha/Sucr:H + > Malt/ α-met-gluc: H* Maltose	High Low	High levels of expression of this gene during wort fermentation	[22]
	Gal2	Glu > Gal	Mod	Xyl/Ara	low			Induces by presence of galactose. Repressed at high glucose concentrations	[20]
	Mal11, 61					Maltose > Turanose	High	MAL loci of constitutive expression and induced by presence of maltose.	[21]
	Mph2, 3					Maltose/ Maltotriose			[22]
	Irt1–2					Inositol	Low	Repression by the presence of glucose	[26]
	Stl1: H+					Glycerol		Active when found in a system with nonfermentable carbon sources, inducible in saline conditions, transient expression, inactive in the presence of glucose	[24]

Species	Transporter	Hexoses	Km	Pentoses	Km	Other	Km	Regulation	Reference
	Fps1					Glycerol efflux		Form protein channel, essential in maintaining the balance in changes hypoosmotic	[23]
	Fsy1 EC1118	Fru Glu	High Low					Repressed by high concentrations of glucose or fructose and was highly expressed on ethanol as the sole carbon source	[27]
	Gup1 / 2					Glycerol		They allow medium growth with glycerol as the sole carbon source and stabilize cell under salinity conditions, Membrane-bound O-acyl transferases family.	[28]
Scheffersomyces stipitis	Xut1/3	Glu/Fru	NR	Xyl	High			Preference for xylose over glucose but moderate transport efficiency	[29]
	Qup2	Glu/Fru	NR	Xyl	NR				
	Hxt2.6	Glu/Fru	NR	Xyl	NR				
	Arby			Ara				Repression by the presence of glucose	[20]
	Gt1					Glycerol	NR	Is active when the medium contains ethanol and absence of sugars	[25]
	Sut1	Glu > Fru	High	Xyl	Low			Induced by glucose presence	[9]
	Sut2	Glu	High	Xyl	Low			Constitutive expression under aerobic conditions and are independent of carbon source	
	Sut3	Glu > Fru	High Low	Xyl > Gal	Low			Constitutive expression under aerobic conditions and are independent of carbon source	
Candida intermedia	Gxs1	Glu: H+		Xyl: H+	High			Repression by the presence of glucose	[18]
	Gxf1	Glu	Low	Xyl	Low			Constitutive expression	
Candida albicans	Hgt1/Hgt2	Glu	High					Repression by high glucose levels	[30]
	Hgt7	Glu	NR					Induced by low glucose levels	
	Hgt12	Glu	NR					Induced by low glucose levels	
	Stl1					Glycerol: H+	High	Induced by salt stress	[31]

Species	Transporter	Hexoses	Km	Pentoses	Km	Other	Km	Regulation	Reference
Yarrowia lipolytica	Yht1	Glu/Fru/Man	NR	Xyl	NR			Main hexose transporters Induced by presence of glucose and galactose	[19] [32]
	Yht2	Fru	NR					Detected in stationary phase of growth	
	Yht3	Fru	NR					Detected in stationary phase of growth	
	Yht4	Glu/Fru/Man	NR					Main hexose transporters Induced by presence of glucose and galactose	
	Yht6			Xyl	NR			Detected in stationary phase of growth	
Schizosaccharomyces pombe	Ght1	Glu > Fru: H+ / Fru	High						[12]
	Ght2	Glu: H+ Fru: H+	Low High						
	Ght3	Glu: H+	High			Gluconate		Transitory expression Gluconate transport inhibited by glucose presence	
	Ght5	Glu > Fru	High Low					Constitutive expression in different carbon sources	
	Ght6	Fru > Glu: H+	High	Fru	High			Constitutive expression in different carbon sources	
	Sut1					Mal > Suc: H+	High	No specific induction. Glucose repression	[33]
Kluyveromyces lactis	Hgt1	Glu > Gal	High					Constitutive expression with 26–31% identity with Hxt in *S. cerevisiae.*	[17]
	Ftr1	Fru: H+	High					Induced by presence of glucose, fructose and galactose	[34]
	Lac2	Gal	Low			Lac	High		[17]
	Rag1	Glu	Low	Fru	NR			Induced by high levels of glucose, fructose and other sugars. Repressed by absence of glucose	[35]
	Kht1	Glu	Low					Induced by high levels of glucose, fructose and other sugars	

Species	Transporter	Hexoses	Km	Pentoses	Km	Other	Km	Regulation	Reference
	Kht2	Glu	High					Induced by low glucose levels	
								Repressed by high glucose levels	
	Kht3	GFal	NR						[36]
Debaryomyces hansenii	Xylh			Xyl: H+	High			Induced by the presence of xylose	[37]
Saccharomyces pastorianus and bayanus	Fsy1	Glu Fru: H+	High			Sorbose	Low	Induced by low Fru levels	[14]
Zygosaccharomyces rouxii	Ffz1	Fru: H+	High					Induced by High levels of Fru	[38]
	Ffz2	Fru	Low						
	Fsy1	Glu Fru: H+	Low High					Induced by low Fru levels	[16]

NR: No reported, Glu: Do not glucose transport, Mod: Affinity modulated by substrate concentrations.
*Modified of Leandro 2009.

Table 1. Yeast transporter with affinity substrate ratios and expression regulation.

http://dx.doi.org/10.5772/intechopen.70783

Once synthesized, they must have to be transported to different compartmentalized organelles such as Nucleus, Endoplasmic Reticulum (ER), Mitochondria (Mt), Golgi bodies, Vacuoles and Peroxisomes [39–41]. These compartmentalized organelles are constituted by multiple sites like outer membrane, intermembrane space, inner membrane and matrix as shown in **Figure 3**. The proteins should be transported to all specified sites of organelle(s) and across the wall (cell wall) to the extracellular medium [39–41]. The order of events that leads the protein to get transported are protein recognition and its subsequent translocation into the organelle. Despite the organelle specific transport, multiple steps of protein transport are briefly generalized here.

3.1. Signal sequence

Most proteins synthesized in the cytosol are mostly precursors or preproteins carrying signal sequences [39]. The signal sequences, present in each protein molecule, are organelle specific. They can be found either at the N-terminal or C-terminal ends of proteins [39–41]. The signal sequence has three conserved general domains: A N-terminal region that varies widely in length, but typically, contains amino acids which contribute a net positive charge: a central hydrophobic region made up of seven to 16 amino acids; followed by a signal cleavage site (**Figure 4**). For instance, Mt. preproteins are rich in positively charged amino acids, arginine and lysine, and hydroxyl bearing ones, serine and threonine. In nuclear preproteins, the sequence region of first 10–90 N-terminal residues, exhibiting a high composition of arginine and near absence of

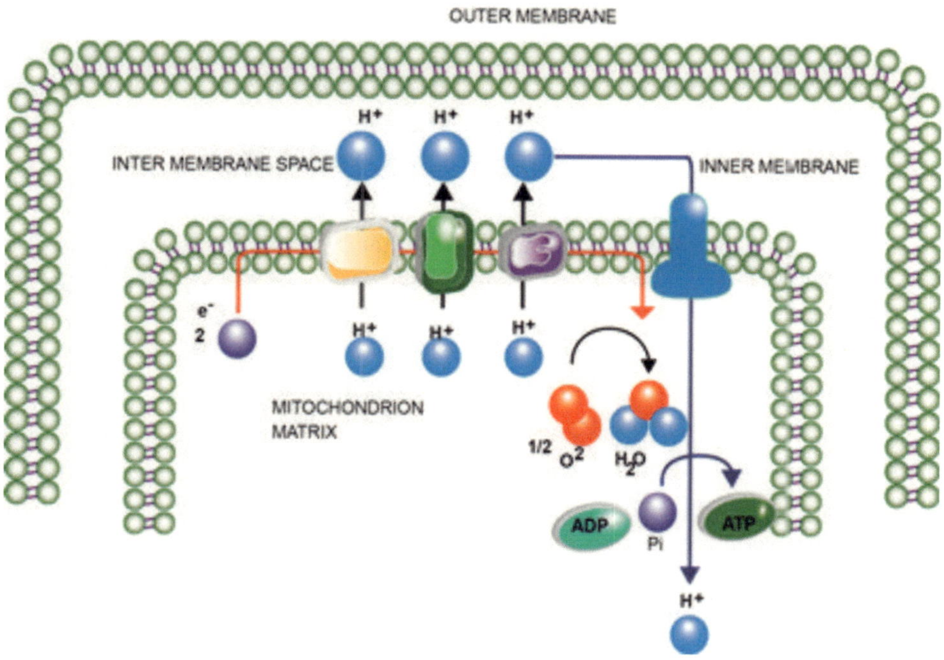

Figure 3. Compartmentalization of organelles like nucleus, endoplasmic reticulum, mitochondria, peroxisomes into multiple layers include outer membrane, intermembrane space, inner membrane and matrix. A mitochondria layers example is given here.

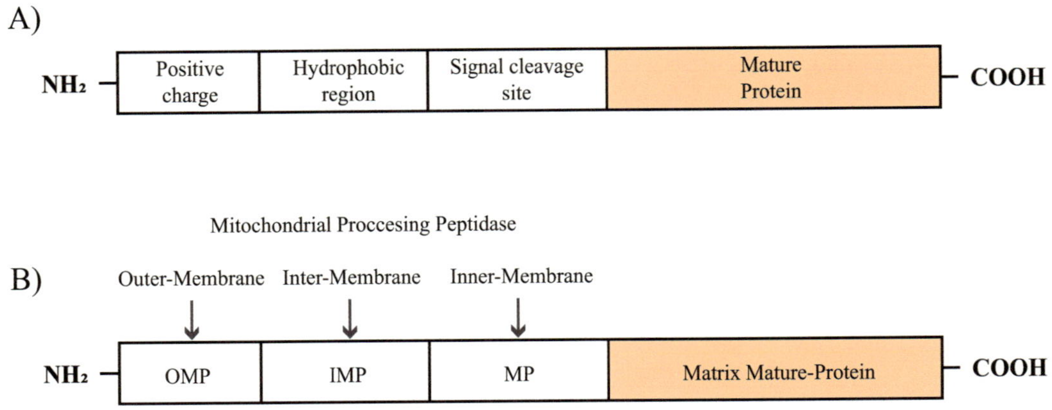

Figure 4. Representation of signal peptide regions present in the various preproteins. (a). The classic preproteins that are entering ER organelle consists of three separate regions in signal peptide include a segment of positively charged amino acids followed by a stretch of hydrophobic amino acids. A protease cleavage site is found next to the mature protein segment. (b). More similarly, mitochondrial matrix proteins contain three peptide regions that corresponds to specific peptidases located on the outer membrane, intermembrane space and inner membrane. Each region is highlighted by arrow. The successive cleavage of peptide regions at respective sites moves the proteins across the membrane to reach matrix.

negatively charged residues, is considered as signal peptide [42]. Regarding membrane proteins, the targeting signals have so far only been identified for a small subset of proteins [43]. In general, non-membranous proteins carry signal peptides at N-terminal, whereas signal peptides are located at the carboxyl termini of membranous proteins [43]. Additional signal sequences found in the proteins conceive multiple entries across the membrane layers of organelles. The example is shown in **Figure 4**, where Mt. luminal proteins contain three signal sequences as follows: (1). A N-terminal protein signal required to gain access into organelle, (2). A stretch of amino acids signalizing the intermembrane space and (3). The mature part of the precursor protein signal that allows the protein to locate themselves into the Mt. lumen [44, 45].

As proteins contain unique signals to each organelle, various bioinformatics databases are developed to facilitate the search process of signals in the proteins. The databases are listed at the end of the book chapter. The enlisted bioinformatic databases will assist the researchers to study and explore the signal peptides appropriate for the organelles of interest.

3.2. Protein recognition and entry into organelle

The signal sequences present in the protein molecules are recognized by signal receptors or signal recognition particles and outer-membrane translocases [42, 44, 45]. They are usually found either in the cytosol or on the membrane of the organelles'. Pex5p is a remarkable example of cytoplasmic receptor protein [46]. Some examples of membranous receptors are exportins and importins (nuclei), translocase outer-membrane complex (Tom70; Mt) [44, 45]. The receptor always function by coupling with other accessory proteins to import and export proteins. For instance, Tom 70 binds to a subset of mitochondrial precursor proteins, with Tom70, are Tom22, Tom5, Tom6, Tom7, Tom20 and Tom70 [44, 45]. These binding partners cooperate and facilitate the targeting of mitochondria proteins. Usually, receptors contain binding sites for signal sequence in the precursor proteins. After gaining access to the organelle specific receptors, precursor proteins are either further processed and deposited into the respective compartments, or translocated directly through the membrane pore complexes [39–41]. In the case of processing precursors, the function of multiple peptidases locating in the respective compartment is required. Especially, in the transport of Mt. luminal proteins, three peptidases: Mitochondrial processing peptidase, Mitochondrial intermediate peptidase, Mitochondrial inner membrane peptidase and their complex proteins are involved to translocate protein from the outer - membrane to the matrix [44, 45] (**Figure 3**).

In the following section, we account on the examples of two typical protein transport systems based on the presence (peroxisome protein) and absence of signal sequences (vesicle-associated protein).

3.3. Transport of peroxisome proteins

Peroxisomes are ubiquitous eukaryotic cell organelles that compartmentalize a large variety of oxidative metabolic reactions. Peroxisome proteins play essential roles in glycolate recycling, amino acid biosynthesis and in fatty acid degradation. Since, it does not contain any genetic material, all the peroxisome proteins are encoded in the nuclear genome. Two types of Peroxisome transport sequence (PTS) have been discovered: type I (PTS1) and type II (PTS2) to translocate proteins from cytoplasm [46, 47]. Some of the identified peroxisome signal peptides are listed in **Table 2**. The PTS1 is found in most of the peroxisome matrix proteins and is located at the C-terminus as a tripeptide SKL20. It generally fits the consensus sequence (S/A/C)-(K/R/H)-(L/M). The PTS2 is a conserved sequence which is located near the N-terminus of a protein and is comprised in some species within a pre-sequence that is cleaved off after import into the peroxisomal matrix. Sequence comparisons showed the conserved nonapeptide of PTS2 as (R/K)-(L/V/I)-X5-(H/Q)-(L/A/F). Some proteins which do not contain neither a PTS1 nor a PTS2 have been identified and well known examples are acyl-CoA oxidase, catalase from *S. cerevisiae* and *Y. lipolytica*, the alcohol oxidase from *Hansenula polymorpha* [48].

Pex5p protein, the cytoplasmic receptor, shuttles between a soluble form and an integral membrane-bound form [46, 49, 50]. They guide free-ribosomal-synthesized peroxisome proteins to translocate across the peroxisome membrane to matrix. It has been characterized that this protein has the capacity to translocate folded, and even oligomeric proteins. The C-terminal domain comprises of seven tetratricopeptide (TPR) repeats, in which 1–3 and 5–7 TPRs adopt extended conformation to link other three TPRs [49]. This conformation produces a funnel shaped binding site for the proteins containing PTS1 signal sequence. Once the receptor recognizes the cargo in the cytosol, a set of proteins Pex13p, Pex14p, Pex17p associate to it forming a docking

Yeasts	Protein	Sequence
C. tropicalis	Catalase	ILELSPRK
S. cerevisiae	Catalase	ELSSNSLF
C. tropicalis	Acyl-CoA oxidase	EYAAILSK
H. polymorpha	Dihydroxyacetone synthase	NHDKVNKL
C. tropicalis	Trifunctional enzyme	LVGDLAKI
S. cerevisiae	Trifunctional enzyme	LSQAKSKL

Table 2. List of peroxisome protein signal sequences.

complex [46, 50]. This establishes a possible link to cargo-receptor complex with peroxisome membrane. At the peroxisome membrane, Pex5p would act as intrinsic membrane protein forming a stable complex with the docking proteins. This complex is shown to exhibit the main conductance of a pore with 3.8 nm in diameter [46, 50]. Also, they can transiently expand to more than 9 nm, when they are importing large oligomeric cargo proteins. The formed pore might at some stage import and translocate the proteins to the lumen [46, 50]. After the luminal protein is released, Pex5p is recycled and translocated to the cytosol by an ATP dependent ubiquitination machinery [46, 49, 50]. In summary, in the cytosol, Pex5p functions as PTS1-receptor in cargo recognition and at the peroxisome membrane where it contributes to pore formation and presumably translocation (**Figure 5**).

3.4. Transport mechanism of a transmembrane protein, Snc1p/2p

Here, we give an example of transport of a transmembrane protein associated to vesicles (discussed below in the following section). Synaptobrevin (Snc1p/Snc2p) is a soluble N-ethylmaleimide-sensitive factor attachment protein receptor (SNARE) transmembrane protein. These proteins bind onto vesicles and interact with t-SNARE proteins on the plasma membrane, by which they provide specificity for the targeting and fusion of vesicles with the plasma membrane [51]. It consists of a variable N-terminal domain, a central coiled-coil domain, and, in most cases, of a single C-terminal transmembrane domain (TMD) that is thought to be α-helical. The conserved region in the SNARE proteins was predicted to contain two amphipathic alpha helices [51]. Helix 1, from 39 to 53, is unusually hydrophobic and Helix

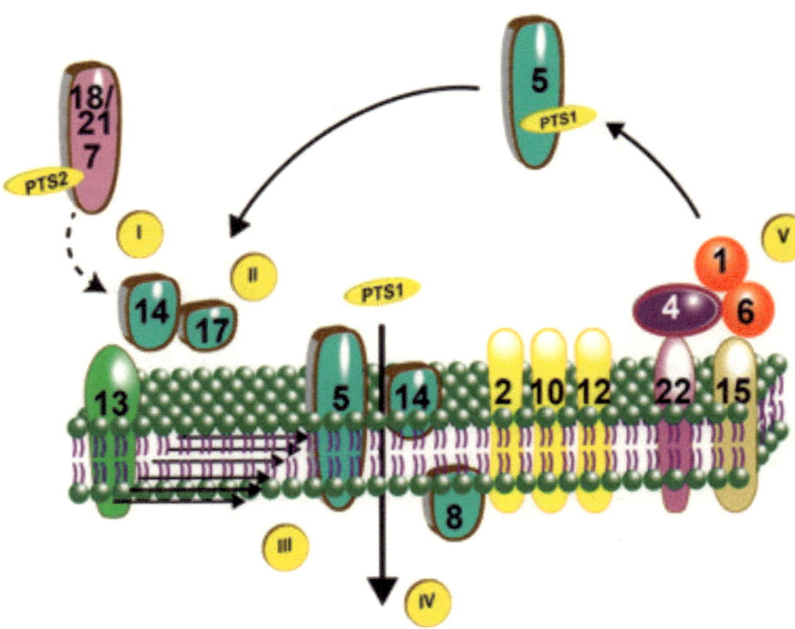

Figure 5. Outline of transportation of matrix proteins of peroxisome mediated by Pex5p and its associated proteins. The protein transport involves five steps: (I) cargo-receptor recognition (II) docking of cargo-receptor complex to the membrane (III) translocation of cargo into matrix across membrane (IV) disassembly and (V) recycling of receptor to cytosol (modified from [50]).

2, from 60 to 88, predicted to be interacted with other hydrophobic segments of membrane proteins t-SNAREs (Syntaxin) during the fusion of vesicles. Other than helices, it carries a variable domain in the N-terminal, a carboxy trans-membrane domain (TMD) region of 96–110 amino acids is usually hydrophobic and some amino acids present intravesicular in vesicles [51].

Just like the other class of membrane proteins, it lacks a signal sequence and contains a single hydrophobic segment close to their C-terminus, leaving most of the polypeptide chain in the cytoplasm (tail-anchored) [43]. The initial targeting of these proteins to the ER is mediated by hydrophobic signal sequences, which are recognized during translation by the signal recognition particle. This hydrophobic stretch near the C termini of membranous protein do not bind to signal recognition particles and are inserted into membranes post-translationally. Once after getting entry into ER, it wasn't clear about the regions responsible in targeting them to secretory vesicles. Deletion and mutational studies were made in the SNARE proteins to investigate the region possessing the ability to target it (**Table 3**). From the targeting studies of Grote et al. [52] and Gerst [53], it was clear that in the absence of helical loops, it is not possible to target the Snc proteins onto secretory vesicles. Thus, deletion or gross substitutions in either of the predicted H1or H2 segments result either in the loss of targeting or in a complete loss of functions. This shows that conserved amphipathic alpha helical region (32–85) is essential for the confinement of snare proteins.

Regions deleted		Effects		Reference
		Presence	Absence	
VAMP	2–30	++		[52]
	2–60		––	
	31–38		––	
	41–50		––	
	61–70	++		
	71–80	++		
Snc1	2–27	++		[53]
	31–50		––	
	51–82	N.D.		
VAMP	1–90		––	
Snc1	91–116		––	
Snc1	1–65	++		
VAMP	65–84			
Snc1	85–116			

++ confers the targeting of SNARE proteins.
–– confers the non-targeting of SNARE proteins.
VAMP—vesicles associated membrane protein/ortholog of Snc1.

Table 3. Deletion mutational study reveals the regions required for the targeting of Snc1 and its ortholog proteins.

In the other hand, deletions of both variable domain and transmembrane domain do not produce a more deleterious effect in the fusion of vesicles. That is, their localization onto vesicles is not affected by these mutations [52–54]. These results substantiate that the TMD of Snc protein is tuned to conduct its delivery into ER, while the helices take it over from ER to Golgi. Besides, the targeting of SNAREs to vesicles, TMD plays a key role in their sorting and fine tunes their distribution within the secretory pathway. That is, TMD sorts Sncp proteins and let them to undergo a dynamic cycle of transport to and retrieval from the plasma membrane to vesicles. Thus, it is understood that TMD serves both, as a key factor in the membrane distribution and as the targeting signal for initial insertion of protein to ER domain. Taking together, it was concluded that the sequence-specific information present in the membrane proteins is important for the respective localization to specific organelles and its subsequent protein function.

4. Vesicular transport

Despite the appreciable functionality of various transporters and protein machinery, there is another existing sophisticated source to transport materials across the walls. They are "naturally existing liposomes" which are made up of an outer hydrophobic lipid bi-layer and an inner aqueous hydrophilic core. Two vesicle types depending on their localization: intracellular and extracellular vesicles are identified and extensively studied in the literature. This section briefly describes the role of such vesicles in the transport of biological materials in yeast organisms.

4.1. Intracellular vesicles

In *S. cerevisiae*, two types of intracellular vesicles: - early secretory and post secretory vesicles are involved to transport cargos (proteins) [39–41]. Early secretory vesicles (ESVs), derived from ER membrane, carry cargoes to Golgi complex, where at this stage post translational modifications such as glycosylation, mannosylation, acetylation, methylation, phosphorylation and acylation are done [41]. While post secretory vesicles (PSVs), shed from trans-Golgi membrane, transport selective cargos destined to extracellular medium and plasma membrane [40]. First, Golgi complex cargoes, secretion and cargoes of other organelles requiring post translational modification are selected via the signal peptide. Later, they are transferred from the ER and packaged as cargo into COPII coated vesicles (ESVs) bound for the stacks of the Golgi complex [39, 41]. The ER vesicles fuse with the cis-Golgi membrane to deposit the cargo into the Golgi complex. The deposited proteins are post translationally modified according to their functional requirements [39, 40]. Then, the proteins for secretion and plasma membrane, that are sorted away from the rest of the cargo, pass into the trans-Golgi network (TGN) and are packed into clathrin coated vesicles, which are called early as PSVs [39].

The PSVs move vectorially towards sites of polarized growth (the bud and mother/daughter neck). They move to arrive at the target membrane dock and subsequently fuse to transfer their contents to extracellular medium [39, 55]. This complete process is termed as polarized exocytosis. It consists of at least three stages. First, PSVs are targeted to the vicinity of designated plasma membrane domains via microtubule- and/or actin-based transport systems [55, 56]. Second, after vesicles arrive at their sites of active exocytosis, where a exocyst complex consisting of eight components: Sec3, Sec5, Sec6, Sec8, Sec10, Sec15, Exo70, and Exo84 mediate the targeting and tethering of PSVs for subsequent membrane fusion [56–58]. Two proteins Sec15p and Sec10p bridge Sec4p, a Rab Gtpase, to other exocyst components. On the plasma membrane, Sec3 and Exo70 interact with PIP2 and with other family members of Rho Gtpases (Cdc42, Rho1p) [58]. Finally, the fusion between PSVs and plasma membrane takes place allowing the secretion of vesicle contents and the incorporation of membrane proteins at specific plasma membrane domains. This specific fusion event is mediated by interaction of proteins present in PSVs membrane (v-SNAREs, snc1p/2p) (SNARE, soluble N-ethylmaleimide-sensitive fusion attachment protein receptors) and plasma-membrane (t-SNAREs; sso1p/2p) [51].

Wild type *S. cerevisiae* strains generate PGVs from 50 to 70 nm in diameter. In contrast, some mutant strains deficient in vesicular transport accumulates PSVs within the cell in different size ranges. For example, exo70–35 and exo70–38 mutant cells accumulated PSVs from were 80–100 nm in diameter [58]. Forsmark et al. [59] have determined the protein composition of PGVs obtained from the sec6–4 and sro7 mutant strains for isolation. The protein content identified are mainly involved in vesicle transport, molecules transportation, metabolism of carbohydrates and protein biosynthesis and degradation. Major dominant lipids constituents of membrane are phosphatidylcholine (PtdCho), phosphatidylethanolamine (PtdEtn) and phosphatidylinositol (PtdIns). It has been speculated that the above mentioned mutants serve as an outstanding source of vesicles for lateral biotechnological applications.

4.2. Exosomes or extracellular vesicles

In all the three kingdoms of life, Exosomes or Extracellular Vesicles (EVs) are one of the most protective sources of conducting trans-cell wall transfer of macromolecules to the recipient cells. EVs (Ø 50–120 nm) are secreted from cells as membranous vesicular organelles by a wide variety of cells, from lower to higher eukaryotic organisms, i.e., from fungi to mammals [60, 61]. Unlike intracellular vesicles, they act as extracellular carriers of proteins and/or nucleic acids, particularly microRNAs and mRNAs, between cells and serve as shuttle vectors and mediators of intercellular communication, immune responses, and antigen presentation [60]. The biogenesis of exosomes begins in the last stage of endocytosis, during which the endocytic membrane undergoes budding to form intraluminal vesicles (ILVs). The accumulated ILVs within the original endocytic membrane, at this stage, is named the multi-vesicular body. These bodies, then fuse with either lysosomes for degradation or the plasma membrane for extracellular release of ILVs, i.e., exosomes or EVs. EVs are released from cells, either constitutively or upon activation of a secretory pathway [60]. The machinery involved in the biogenesis of exosomes varies in different cell types [60, 61]; however, in most cells, the ESCRT (endosomal sorting complex required for transport) machinery plays a major role in EVs biogenesis [60, 61]. The roles of both the ESCRT-dependent and -independent mechanisms in exosome biogenesis remain largely unknown and are yet to be fully elucidated. Similarly, the mechanisms underlying the packaging of cargo into exosomes and the transport of these exosomes across cellular membranes have been described both *in vivo* and *in vitro*, but remain to be fully elucidated [60, 61].

EVs have been conserved and distributed widely in many different fungal species, including yeast cells and hyphae [23]. Pathogenic fungus and opportunistic fungus are the well-recognized candidates for the release of EVs. Some of the examples are as follows: *Paracoccidioides brasiliensis, Sporothrix schenckii, Candida albicans, Candida parapsilosis, Malassezia sympodialis, Histoplasma capsulatum, Cryptococcus neoformans, Malassezia sympodialis* [61]. Non-pathogenic *S. cerevisiae* mutants (sec4–2, sec6–4, sec4–8, sec23–1, exo70–35, and exo70–38) have also been demonstrated to excrete EVs in the extracellular medium [62]. Several EV proteome studies revealed the presence of multiple organelle specific proteins which are derived from the cytoplasm, plasma membrane, mitochondrial, vacuolar and even nuclear proteins. Sterols, phospholipids and pigments are also present in the EVs. Quite recently, the presence of small RNAs in the fungal EVs was addressed [63].

4.2.1. Diverse roles of transport cargoes of EVs

The EVs derived from pathogenic fungus are natural born carriers of cargo responsible for fungal pathogenesis. Several components of fungal EVs are potent elicitors of immunological activities [64]. For instance, the very common protein HSP60 carried by EVs acts as immunogen and induces protective antibodies [65]. The main virulence factor of EVs derived from *Cryptococcus neoformans* is a polysaccharide capsule coating glucuronoxylomannan, which activates immune-suppressive and anti-phagocytic properties [65]. The incubation of cryptococcal vesicles with murine macrophages induced high levels of extracellular tumor necrosis factor alpha (TNF-α), interleukin-10 (IL-10), and transforming growth factor β (TGF-β) [65]. *M. sympodialis* releases extracellular vesicles carrying allergen were shown to induce IL-4 and TNF-α responses in PBMC patients [66]. The immunogenic galactosyl epitopes distributed on the surface of large EVs of *Paracoccidioides brasilensis* generated robust immune response in the paramycoccidiodomycosis patients [67]. They bind to host lectins and induce immunological type 2 suppressive response. Also, *C. albicans* EVs stimulated dendritic cells (DCs) to produce IL-12p40, IL-10, and TNF-α, and induced upregulation of CD86 and MHC-II [68]. The change in protein composition of THP-1 macrophage-derived EVs was studied during the interaction with *C. albicans* [69]. This study revealed the changes in the abundance of proteins relating to immune response, signaling, or cytoskeletal reorganization. The interaction significantly increased the secretion of proinflammatory cytokines and the candidacidal activity. More likely, the treatment of *H. capsulatum* cells with monoclonal antibodies (MAbs) affected the sizes, enzymatic contents, and proteomic profiles of the vesicles released by fungal cells [70]. The coating of Hc with cryptococcal glycans (Cn-gly) resulted in higher pulmonary fungal burden in co-infected animals relative to control. Co-cultivation or addition of Cn-gly resulted in enhanced pellicle formation with a hybrid polysaccharide matrix with higher reactivity to GXM mAbs [71].

5. Methods to determine the secretion-proteins across yeast wall

Numerous established techniques are already available in literature to detect, characterize and demonstrate the phenomenon of secreting proteins, towards the extracellular medium, across yeasts wall. At cellular level, usually, the proteins destined to secretion are always preserved intact into the secretory vesicles of yeasts. Taking advantage of this nature, many fluorescent methods detect the proteins presence througth the secretion route in cells by fluorescence. The availability of several fluorescent proteins (FPs): Green-FP, Red-FP, Yellow-FP and Blue-FP has made the detection process simple and effective [72]. To this end, tagging proteins of interest with FPs, using genetic engineering techniques, will come handy and serve the purpose of locating them into the cells. In the other hand, immunofluorescent technique makes use of antibodies to demonstrate the integrity of secretion proteins inside vesicles [53]. For this purpose, various temperature sensitive sec-mutant strains, with the ability to accumulate vesicles, are highly recommended [73].

Once the proteins are secreted outside, they can be characterized by molecular techniques like SDS-PAGE and Western blotting to identify specifically the proteins of interest in the extracellular medium [53, 74]. By other hand enzyme activity studies are suitably advantageous to determine the proper functioning of the secreted protein The design of such experiments generally varies with respect to the enzymes and must be handled appropriately and the experiments can be performed either by using the whole extracellular medium containing secreted proteins or by using the purified proteins of interest (see protein purification section in applications below). Combining all together, we conclude that one of the abovementioned techniques could be suitable for realizing adequate studies on the proteins secretion.

6. Applications in biotechnology

This is an overview of the main trends reported within the last years in current research on applications related to transport proteins in some yeast, which has not yet discussed in detail. Major advances, of the role of different biological transporters in *S. cerevisiae* are focused in carbohydrates related to obtain value-added bioproducts. Mainly discuss the expression of carbohydrates transporters in yeast are focused to improve different substrates and in the modification of specific aminoacides into transporters to regulate the affinity, order to alleviate transport inhibition by sugar concentration. The capacity to co-transport glucose and xylose into yeast has remained a technical challenge in the field [11, 29]. Due to the lack of an endogenous xylose transporter in *Saccharomyces cerevisiae*, the xylose uptake depends on transporter engineering to increase transport rates avoiding glucose-based inhibition, thus enhancing the potential of using lignocellulosic biomass as a feedstock for yeast [11, 29]. Besides of to the generation of fuels, the production of value-added chemicals from renewable biomass has been widely studied. According to [75], *S. cerevisiae* could be exploited for the production of other non-ethanol fuels and chemicals from byproducts through metabolic engineering expressing specific sugar transporter. Some other efforts to use *S. cerevisiae* strains as a cell factory to obtain valued-added products which no involves the use of genetic tools, but metabolic activators [76] and ultrasound [77].

Some other works involves the study of trafficking mechanisms of small and large compounds to regulate biosynthesis of appreciated biochemical products. Also, mitochondrial transport mechanisms are relevant due to its use in future comparative studies aiding explorations of human mitochondrial diseases and to improve biochemical process. Because energy is a fundamental enabler of the economy, energy security and environmental safety are two major issues in the current world that have boosted the demand for an alternative and eco-friendly energy source.

6.1. Protein purification mediated by heterologous expression

Using genetic engineering techniques, recombinant proteins can be synthesized in anyone of three compartments of heterologous hosts: cytoplasm, periplasm and the extracellular medium. The natural ability of secreting proteins is captivated by many researchers as a medium for the large-scale industrial production of foreign proteins and simplifying downstream processes [78]. The secretory expression requires a simple tagging of recombinant proteins of interest with three essential components: (1). A signal peptide sequence targeting secretion, followed by (2) a purification tag and (3) a protease cleavage site [78–80]. Some of the examples of these three essential components are enlisted and the recommendable design of a gene fusion cassette for recombinant protein secretion is shown in **Figure 6**. The expression of this gene fusion cassette in the following hosts enables the secretion of protein towards extracellular medium. The purification tag serves as an anchor and allows the recombinant protein to separate from rest of the media culture, which is subsequently recovered by using protease enzyme [80]. Some of the valuable hosts as recommended by Food and Drug Administration are *S. cerevisiae, P. pastoris, Y. lipolytica, K. lactis,* and *H. polymorpha* [78]. Though technology ages 3 decades, the growth and value of applications are still increasing with respective to the demand. Some of the recent heterologous expression and secretion of proteins of biotechnological interest are presented in the **Table 4**. It is important to note that the α-MF signal sequence has proven to be most effective in directing protein through the secretory pathway in host organisms [81, 84–91, 93, 94]. Such expression and purification of recombinant proteins are widely applied in the industries of textile, food processing, therapeutic applications. In the other hand, the natural ability of yeast hosts to provide, post translational modifications was highly utilized to express, modify and further secrete eukaryotic proteins, especially for therapeutic applications, in the extracellular medium [99]. The expression and secretion of full length IgGs, insulin, glucagon, growth hormone, in yeast hosts is a proof-of-concept in this context [52]. The glycolate form of antibodies and human glycoproteins with fully complex terminally sialylated N-glycans were also synthesized in the engineered *Scchefersomyces stipitis* (formerly *P. pastoris*) [97, 98].

6.2. Peroxisome production of valuable bioproducts

Here, we highlight the use of signal peptides and transporter system of Peroxisome for the synthesis of valuable bioproducts. Mostly, researchers took advantage of the active fatty acid

http://dx.doi.org/10.5772/intechopen.70783

(a)

Proteins	Signal Peptide
Galactosidase	MFAFYFLTACISLKGVFG:VSPSYNGLGL
Acid phosphatase	MFKSVVYSILAASLANA:GTIPLGKLAD
Carboxypeptidase Y	MKAFTSLLCGLGLSTTLAKA:ISLQRPL
Invertase	MKIYHIFSVCYLITLCAAATTAREEFF
Mating factor a-1	MLLQAFLFLLAGFAAKISA:SMTNETSDRP
Glucoamylase	MVGLKNPYTHTMQRPFLLAYLVLSLLF NSALGFPTALVPRGS
Mating factor alpha-2	MKFISTFLTFILAAVSVTASSDEDIAQVPA

(b)

Purification Tags

His-tag or His6
Myc
Glutathione S-transferase (GST),
Maltose binding protein
Calmodulin binding peptide
Intein-Chitin binding domain
Streptavidin/Biotin tags
FLAG
Halo
Small ubiquitin-like modifier

(c)

Protease cleavage site

Chymotrypsin
Thrombin
Trypsin
Enterokinase
TEV protease
Enteropeptidase
3C protease
Carboxypeptidase
Aminopeptidase
Factor Xa

(d)

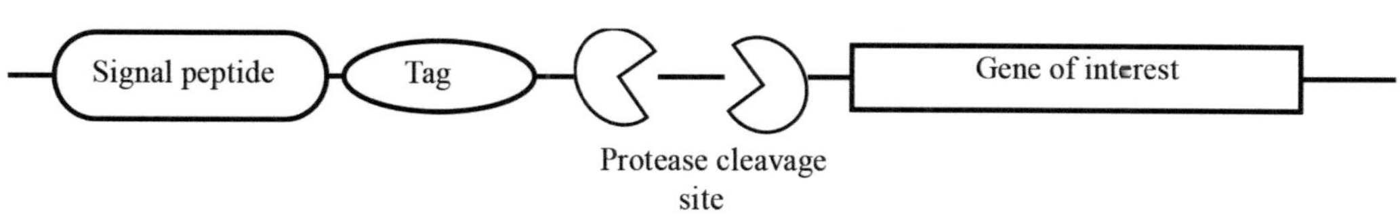

Figure 6. Overview of genetic elements used for the recombinant secretion of proteins towards extracellular medium (a, b, and c). The most significant elements are signal peptide sequence, protein tags and protease cleavage sites. Some examples of well-recognized and highly used components are listed. (d). The design of a gene construct that is in practice and essential for the heterologous expression and secretion of recombinant proteins in yeast hosts in presented.

pathways and PTS1 signals to generate polyhydroxyalkanoates (bioplastics) and biofuels (fatty-acid-derived fatty alcohols, alkanes and olefins) [99–106]. From literature, a simple modification of polyhydroxyalkanoate synthase with PTS was sufficient for targeting and

Organism	Protein	Signal	Applications	Reference
S. cerevisiae				
	Human β-defensin-2, (hBD2)	MFα1 (mating factor alpha) leader	Antimicrobial activity	[81]
	Beta glucosidase	Sed1, glucoamylase, alpha mating leader	Cellulolytic activity	[82]
	endoglucanase II			
	Cel3A	Native secretion signal	Lignocellulosic	[83]
	Cel7A		ethanol production	
	Cel5A			
Pichias pastoris				
	Trx-HPV16-L2 immunogen	alpha-factor signal peptide	Vaccine	[84]
	Horseradish peroxidase	MATα prepro secretion signal		[85]
	Candida antartica lipase			
	Human Pro-relaxin L2	alpha-factor signal peptide	Therapeutic applications	[86]
	FSL2, Lipase	*S. cerevisiae* α-factor signal sequence	Lipolytic activity	[87]
	Endo-polygalacturonase	alpha-factor signal peptide	Textile scouring	[88]
	Camel Hepcidin	*S. cerevisiae* α-factor signal sequence	Antimicrobial activity, Hormone	[89]
	Human anti-αIIbβ3 antibody	alpha-factor signal peptide	Atheroma Targeting	[90]
	Subtilisin QK	alpha-factor signal peptide	Thrombolytic activity	[91]
Yarrowia lipolytica				
	Glucoamylase	preLip2, preXpr2, and preSuc2	Starch degradation	[92]
	Xylanase			
Kluyveromyces lactis				
	Fructosyltransferase	alpha-factor signal peptide	Hypocaloric sweeteners	[93]
	Arylsulfatase	alpha-factor signal peptide	Milk processing	[94]
	Interferon-Beta	Glucoamylase signal sequence	Therapeutic applications	[95]

Table 4. Recent heterologous expression and secretion of recombinant proteins from mid2016–Feb2017.

synthesizing PHAs in peroxisome of *S. cerevisiae* and *P. pastoris* [99–102]. The authors did an addition of carboxyl 34 amino acids from the *Brassica napus* isocitrate lyase to *Pseudomonas aeruginosa* PHAC1 synthase for peroxisome targeting, which was further expressed under the control of the promoter of the *P. pastoris* acyl-CoA oxidase gene [96, 97]. By this expression, PHAs was accumulated as inclusions within the peroxisomes and synthesized up to 1% medium-chain-length PHA per g dry weight was obtained using oleic acid as substrate in the medium. In another study, a medium-chain-length-PHA (mcl-PHA) polymer was synthesized in the cytosol of *S. cerevisiae* utilizing the ß-oxidation intermediates, key peroxisome proteins, including Faa2p, Fox1p, and Fox2p, together with PHA synthase [101]. A Pex5p mutant was made in *S. cerevisiae* to retain peroxisome proteins in the cytoplasm. This retention led the peroxisome proteins to take part actively in the generating the mcl-PHA monomers. Accumulated PHA up to approximately 7% of its cell dry weight with a monomeric composition of C12 (3-hydroxydodecanoic acid), C10 (3-hydroxydecanoic acid), C8 (3-hydroxyoctanoic acid), and C6 (3-hydroxyhexanoic acid).

Another effective exploration is targeting synthetic pathways to peroxisomes to produce medium fatty alcohols and long fatty alcohols [103–106]. The targeted expression of fatty acyl-CoA reductase TaFAR to the peroxisome of *S. cerevisiae* has produced medium chain fatty alcohols [103]. The genes Pex7p and acetyl-CoA carboxylase are overexpressed together with targeted TaFAR enzyme in the peroxisome. The coexpression improved the synthesis of decanol, dodecanol, tetradecanol and hexadecanol, which have extensive applications as biofuels and detergents. Another heterologous expression of a fatty acyl-CoA reductase from *Arabidopsis thaliana* in a Pex10p mutant *Y. lipolytica* had produced over 500 mg/L of 1-decanol [104]. Likewise, *Rhodospirillum toruloides* was engineered to express a bifunctional fatty acyl-ACP reductase (FaCoAR) from *Marinobacter aquaeolei* VT8 and produced up to over 8 g/L of C16–C18 fatty alcohols in fed-batch condition using sucrose as carbon source [105]. A recent study has shown enhanced the peroxisome production of fatty alcohols by targeting the FaCoAR enzyme using signal per2 (GGGSAAVKLSQAKSKL) [100]. In the same study, the expression of two FFA based enzymes, *Mycobacterium marinum* carboxylic acid reductase (MmCAR)29 and its activation cofactor-4′-phosphopantetheinyl transferase NpgA from *Aspergillus nidulans* in a Pex31p/Pex32p mutant strain have resulted in the high level of alkane production.

6.3. Vesicles in therapeutic applications

The prime role of intercellular communication has motivated researchers to conceive EVs as potential nano-vehicles for biodelivery applications. Recently in 2016, A patent entitled ¨Compositions and Methods for Yeast Extracellular Vesicles as Delivery Systems, US 20160331686¨ was filed and published [107]. The authors have proposed the use of native and modified EVs from yeasts cells as practical drug delivery vehicles. In the case of modified EVs, an exosomal transmembrane peptide of mammalian origin is immobilized onto the outer membrane of EVs for targeted biodelivery applications. Using these yeast EVs, various therapeutic sources of cargoes: therapeutic RNAs (circular RNAs), autonomously replicating cytoplasmic linear mammalian plasmid (express either therapeutic RNAs or proteins), therapeutic peptides, have been tested for delivery applications. Once the cargo loaded EVs are released from cells, they have been isolated from culture supernatants by either centrifugation or ultra/micro filtration. Authors conducted *in vivo* and *in vitro* studies to study the uptake of EVs by these cells and its effect in the delivery of cargoes. The purified vesicles are recognized by mammalian target cells with the receptors specific for the targeting ligand and take up the vesicles carrying the biologically active therapeutics via endocytosis. Following this inspirational work, we believe that the combination of recombinant DNA techniques and natural loading efficiency of cargoes into EVs would bring potential drug-targeting properties in future.

7. Webserver

Mitochondria

- p://mitf.cbrc.jp/MitoFates/
- MitoII

Subcellular localization program

- PSORT
- TargetP
- NNPSL (neural network-based predictor)
- http://www.signalpeptide.de/
- SignalP 4.1 (http://www.cbs.dtu.dk/services/SignalP/)
- SPdb (http://proline.bic.nus.edu.sg/spdb)

Peroxisome

- PTS1 Predictor - http://mendel.imp.univie.ac.at/mendeljsp/sat/pts1/PTS1predictor.jsp

Peptidase Database

- http://merops.sanger.ac.uk/

Acknowledgements

The authors acknowledge Juan Corona-Hernandez and Alonso Cid Cervantes for their artwork contribution. GKM thank CONACyT (CB-2014-01; 236285 and Fronteras de la Ciencia 2015-1: 016) for financial support. López-vargas acknowledge to scholarship of PNPC-CONACyT program 5353.

Author details

Aline López-Vargas[1], Gurusami Kutralam-Muniasamy[2], Aldo Amaro-Reyes[1], Fermín Pérez-Guevara[2] and Jorge Noel Gracida Rodríguez[1]*

*Address all correspondence to: gracidaj@netscape.net

1 Facultad de Química, Universidad Autónoma de Querétaro, Queretaro Qro, Mexico

2 Biotechnology, Centro de Investigaciones y de Estudios Avanzados del Instiuto Politécnico Nacional, Cd de México, México

References

[1] Kim JH, Roy A, Jouandot D II, Cho KH. The glucose-signaling network in yeast. Biochimica et Biophysica Acta. 2013;**1830**(11):5204-5210. DOI: 10.1016/j.bbagen.2013.07.025

[2] Özcan S, Dover J, Rosenwald AG, Wölfl S, Johnston M. Two glucose transporters in Saccharomyces Cerevisiae are glucose sensors that generate a signal for induction of gene expression. Proceedings of the National Academy of Sciences. 1996;**93**(22):12428-12432

[3] Conrad M, Schothorst J, Nah-Kankipati H, Van-Zeebroeck G, Rubio-Texeira M, Thevelein JM. Nutrient sensing and signaling in the yeast Saccharomyces Cerevisiae. FEMS Microbiology. 2014;**38**(2):254-299. DOI: 10.1111/1574-6976.12065

[4] Roy A, Shin YJ, Cho KH, Kim JH. Mth1 regulates the interaction between the Rgt1 repressor and the Ssn6-Tup1 corepressor complex by modulating PKA-dependent phosphorylation of Rgt1. Molecular Biology of the Cell. 2013;**24**(9):1493-1503. DOI: 10.1091/mbc.E13-01-0047

[5] Lakshmanan J, Mosley AL, Ozcan S. Repression of transcription by Rgt1 in the absence of glucose requires Std1 and Mth1. Current Genetics. 2003;**44**(1):19-25. DOI: 10.1007/s00294-003-0423-2

[6] Sexton JA, Brown V, Johnston M. Regulation of sugar transport and metabolism by the Candida Albicans RgtI transcriptional repressor. Yeast. 2007;**24**(10):847-860. DOI: 10.1002/yea.1514

[7] Ng TS, Mohd-Desa, Sandai D, Chong PP, Than LT. Phylogenetic and transcripts profiling of glucose sensing related genes in Candida Glabrata. Jundishapur Journal of Microbiology. 2015;**8**(11):e25177. DOI: 10.5812/jjm.25177

[8] Cairey-Remonnay A, Deffaud J, Wésolowski-Louvel M, Lemaire M, Soulard A. Glycolysis controls plasma membrane glucose sensors to promote glucose signaling in yeasts. Molecular and Cellular Biology. 2015;**35**(4):747-757. DOI: 10.1128/MCB.00515-14

[9] Leandro MJ, Fonseca C, Gonçalves P. Hexose and pentose transport in ascomycetous yeasts: An overview. FEMS Yeast Research. 2009;**9**(4):511-525. DOI: 10.1111/j.1567-1364.2009.00509

[10] Reifenberger E, Freidel K, Ciriacy M. Identification of novel HXT genes in Saccharomyces Cerevisiae reveals the impact of hexose transporters on glycolytic flux. Molecular Microbiology. 1995;**16**(1):157-167. DOI: 10.1111/j.1365-2958.1995.tb02400.x

[11] Gonçalves DL, Matsushika A, Sales BB, Goshima T, Bon EPS, Stambuk BU. Xylose and xylose/glucose co-fermentation by recombinant Saccharomyces Cerevisiae strains expressing individual hexose transporters. Enzyme and Microbial Technology. 2014;**63**:13-20. DOI: 10.1016/j.enzmictec.2014.05.003

[12] Heiland S, Radovanovic N, Milan H, Winderickx J, Lichtenberg H. Multiple hexose transporters of Schizosaccharomyces pombe. Journal of Bacteriology. 2000;**182**(8):2153-2162

[13] Gonçalves P, Rodrigues de Sousa H, Spencer-Martins I. FSY1, a novel gene encoding a specific fructose/H(+) symporter in the type strain of Saccharomyces Carlsbergensis. Journal Bacteriology. 2000;**182**(19):5628-5630. DOI: 10.1128/JB.182.19.5628-5630.200

[14] Rodrigues de Sousa H, Spencer-Martins I, Gonçalves P. Differential regulation by glucose and fructose of a gene encoding a specific fructose/H+ symporter in Saccharomyces Sensu stricto yeasts. Yeast. 2004;**21**(6):519-530. DOI: 10.1002/yea.1118

[15] Betina S, Goffrini P, Ferrero I, Wésolowski-Louvel M. RAG4 gene encodes a glucose sensor in Kluyveromyces lactis. Genetics. 2001;**158**(2):541-548

[16] Leandro MJ, Cabral S, Prista C, Loureiro-Dias MC, Sychrová H. The high-capacity specific fructose facilitator ZrFfz1 is essential for the Fructophilic behavior of Zygosaccharomyces rouxii CBS 732T. Eukaryotic Cell. 2014;**13**(11):1371-1379. DOI: 10.1128/EC.00137-14

[17] Baruffini E, Goffrini P, Donnini C, Lodi T. Galactose transport inKluyveromyces lactis: Major role of theglucose permease Hgt1. FEMS Yeast Research. 2006;**6**(8):1235-1242. DOI: 10.1111/j.1567-1364.2006.00107.x

[18] Leandro MJ, Gonçalves P, Spencer-Martins I. Two glucose/xylose transporter genes from the yeast Candida Intermedia: First molecular characterization of a yeast xylose–H+ symporter. Biochemical Journal. 2006;**395**(3):543-549. DOI: 10.1042/BJ20051465

[19] Young EM, Tong A, Bui H, Spofford C, Alper HS. Rewiring yeast sugar transporter preference through modifying a conserved protein motif. Proceedings of the National Academy of Sciences USA. 2014;**111**(1):131-136. DOI: 10.1073/pnas.1311970111

[20] Wang C, Shen Y, Zhang Y, Suo F, Hou J, Bao X. Improvement of L-arabinose fermentation by modifying the metabolic pathway and transport in Saccharomyces Cerevisiae. BioMed Research International. 2013;**2013**:461204. DOI: 10.1155/2013/461204

[21] Dietvorst J, Londesborough J, Steensma HY. Maltotriose utilization in lager yeast strains: MTT1 encodes a maltotriose transporter. Yeast. 2005;**22**(10):775-788. DOI: 10.1002/yea.1279

[22] Alves SL Jr, Herberts RA, Hollatz C, Miletti LC, Stambuk BU. Maltose and maltotriose active transport and fermentation by Saccharomyces Cerevisiae. Journal of the American Society of Brewing. Chemistry. 2007;**65**(2):99-104. DOI: 10.1094/ASBCJ-2007-0411-01

[23] Geijer C, Ahmadpour D, Palmgren M, Filipsson C, Medrala-Klein D, Tamás M, Hohmann S, Lindkvist-Petersson K. Yeast Aquaglyceroporins use the transmembrane core to restrict glycerol transport. Journal of Biological Chemistry. 2012;**287**(28):23562-23570. DOI: 10.1074/jbc.M112.353482

[24] Ferreira C, van Voorst F, Martins A, Neves L, Oliveira R, Kielland-Brandt M, Lucas C, Brandt A. A member of the sugar transporter family, Stl1p is the glycerol/H+ symporter in Saccharomyces Cerevisiae. Molecular Biology of Cell. 2005;**16**(4):2068-2076. DOI: 10.1091/mbc.E04-10-0884

[25] Zhan C, Wang S, Sun Y, Dai X, Liu X, Harvey L, McNeil B, Yang Y. Bai Z. The Pichia Pastoris transmembrane protein GT1 is a glycerol transporter and relieves the repression of glycerol on AOX1 expression. FEMS Yeast Research. 2016;**16**(4):fow033. DOI: 10.1093/femsyr/fow033

[26] Schneider S. Inositol transport proteins. Federal European Biochemical Societies Letter. 2015;**589**(10):1049-1058. DOI: 10.1016/j.febslet.2015.03.012

[27] Galeote V, Novo M, Salema-Oom M, Brion C, Valerio E, Gonçalves P, Dequin S. FSY1, a horizontally transferred gene in the Saccharomyces Cerevisiae EC1118 wine yeast strain, encodes a high-affinity fructose/H+ symporter. Microbiology. 2010;**156**(12):3754-3761. DOI: 10.1099/mic.0.041673-0

[28] Neves L, Oliveira R, Lucas C. Yeast orthologues associated with glycerol transport and metabolism. FEMS Yeast Reserch. 2004;**5**(1):51-62. DOI: 10.1016/j.femsyr.2004.06.012

[29] de Sales BB, Scheid B, Gonçalves DL, Knychala MM, Matsushika A, Bon EP, Stambuk BU. Cloning novel sugar transporters from Scheffersomyces (Pichia) stipitis allowing D-xylose fermentation by recombinant Saccharomyces Cerevisiae. Biotechnology Letters. 2015;**37**(10):1973-1982. DOI: 10.1007/s10529-015-1893-2

[30] Fan J, Chaturvedi V, Shen SH. Identification and phyloge-netic analysis of a glucose transporter gene family from thehuman pathogenic yeast Candida Albicans. Journal of Molecular Evolution. 2002;**55**(3):336-346. DOI: 10.1007/s00239-002-2330-4

[31] Kayingo G, Martins A, Andrie R, Neves L, Lucas C, Wong B. A permease encoded by STL1 is required for active glycerol uptake by Candida Albicans. Microbiology. 2009;**155**(5):1547-1557. DOI: 10.1099/mic.0.023457-0

[32] Lazar Z, Neuvéglise C, Rossignol T, Devillers H, Morin N, Robak M, Nicaud JM, Crutz-Le Coq AM. Characterization of hexose transporters in Yarrowia lipolytica reveals new

groups of sugar porters involved in yeast growth. Fungal Genetics and Biology. 2017;**100**:1-12. DOI: 10.1016/j.fgb.2017.01.001

[33] Reinders A, Ward JM. Functional characterization of the alpha-glucoside transporter Sut1p from Schizosaccharomyces pombe, the first fungal homologue of plant sucrose transporters. Molecular Microbiology. 2001;**39**(2):445-454. DOI: 10.1046/j.1365-2958.2001.02237.x

[34] Diezemann A, Boles E. Functional characterization of the Frt1 sugar transporter and of fructose uptake in Kluyveromyces lactis. Current Genetics. 2003;**43**(4):281-288. DOI: 10.1007/s00294-003-0392-5

[35] Milkowski C, Krampe S, Weirich J, Hasse V, Boles E, Breunig KD. Feedback regulation of glucose transporter gene transcription in Kluyveromyces lactis by glucose uptake. Journal Bacteriology. 2001;**183**(18):5223-5229. DOI: 10.1128/JB.183.18.5223-5229.2001

[36] Wiedemuth C, Breunig KD. Role of Snf1p in regulation of intracellular sorting of the lactose and Galactose transporter Lac12p in Kluyveromyces lactis. Eukaryotic Cell. 2005;**4**(4):716-721. DOI: 10.1128/EC.4.4.716-721.2005

[37] Ferreira D, Nobre A, Silva ML, Faria –OF, Tulha J, Ferreiara C, Lucas C. XYLH encodes a xylose/H+ symporter from the highly related yeast species Debaryomyces fabryi and Debaryomyces Hansenii. FEMS Yeast Research. 2013;**13**(7):585-596. DOI: 10.1111/1567-1364.12061

[38] Leandro MJ, Sychrová H, Prista C, Loureiro-Dias MC. The osmotolerant fructophilic yeast Zygosaccharomyces rouxii employs two plasma-membrane fructose uptake systems belonging to a new family of yeast sugar transporters. Microbiology. 2011;**157**(2): 601-608. DOI: 10.1099/mic.0.044446-0

[39] Bonifacino JS, Glick BS. The mechanisms of vesicle budding and fusion. Cell. 2004;**116**(2): 153-166. DOI: 10.1016/S0092-8674(03)01079-1

[40] Gleeson PA. Targeting of proteins to the Golgi apparatus. Histochemistry and cell biology. 1998;**109**(5–6):517-532. DOI: 10.1007/s004180050252

[41] Pfeffer SR, Rothman JE. Biosynthetic protein transport and sorting by the endoplasmic reticulum and Golgi. Annual Review of Biochemistry. 1987;**56**(1):829-852. DOI: 10.1146/annurev.bi.56.070187.004145

[42] Lange A, Mills RE, Lange CJ, Stewart M, Devine SE, et al. Classical nuclear localization signals: Definition, function, and interaction with importin α. Journal of Biological Chemistry. 2007;**282**(8):5101-5105. DOI: 10.1074/jbc.R600026200

[43] Borgese N, Fasana E. Targeting pathways of C-tail-anchored proteins. Biochimica et Biophysica Acta-Biomembranes. 2011;**1808**:937-946. DOI: 10.1016/j.bbamem.2010.07.010

[44] Endo T, Kohda D. Functions of outer membrane receptors in mitochondrial protein import. Biochimica et Biophysica Acta (BBA)-Molecular Cell Research. 2002;**1592**(1):3-14. DOI: 10.1016/S0167-4889(02)00259-8

[45] Endo T, Yamano K. Transport of proteins across or into the mitochondrial outer membrane. Biochimica et Biophysica Acta (BBA)-Molecular Cell Research. 2010;**1803**(6):706-714. DOI: 10.1016/j.bbamcr.2009.11.007

[46] Baker A, Hogg TL, Warriner SL. Peroxisome protein import: A complex journey. Biochemical Society Transactions. 2016;**44**(3):783-789. DOI: 10.1074/jbc.M203254200

[47] Gould SJ, Keller GA, Hosken N, Wilkinson J, Subramani S. A conserved tripeptide sorts proteins to peroxisomes. Journal of Cell Biology. 1989;**108**:1657-1664. DOI: 10.1083/jcb.108.5.1657

[48] Klein AT, van Den Berg M, Bottger G, Tabak HF, Distel B. Saccharomyces Cerevisiae acyl-CoA oxidase follows a novel, non-PTS1, import pathway into peroxisomes that is dependent on Pex5p. Journal of Biological Chemistry. 2002;**277**:25011-25019. DOI: 10.1074/jbc.M203254200

[49] Gatto GJ, Geisbrecht BV, Gould SJ, Berg JM. Peroxisomal targeting signal-1 recognition by the TPR domains of human PEX5. Nature Structural Biology. 2000;**7** 1091-1095. DOI: 10.1038/nsb1002-788b

[50] Rucktäschel R, Girzalsky W, Erdmann R. Protein import machineries of peroxisomes. Biochimica et Biophysica Acta (BBA)-Biomembranes. 2011;**1808**(3):892-900. DOI: 10.1016/j.bbamem.2010.07.020

[51] Gerst JE. SNAREs and SNARE regulators in membrane fusion and exocytosis. Cellular and Molecular Life Sciences. 1999;**55**(5):707-734. DOI: 10.1007/s000180050328

[52] Grote E, Hao JC, Bennett MK, Kelly RB. A targeting signal in VAMP regulating transport to synaptic vesicles. Cell. 1995;**81**:581–589. DOI: 10.1016/0092-8674(95)90079-9

[53] Gerst JE. Conserved helical segments on yeast homologs of the synaptobrevin VAMP family of V-snare mediate exocytic functions. Journal of Biological Chemistry. 1997;**272**: 16591-16598. DOI: 10.1074/jbc.272.26.16591

[54] Protopopov V, Govindan B, Novick P, Gerst JE. Homologs of the Synaptobrevin VAMP family of synaptic vesicles protein function on the late secretory pathway in Saccharomyces Cerevisiae. Cell. 1993;**74**(5):855-861. DOI: 10.1016/0092-8674(93)90465-3

[55] Pruyne D, Bretscher A. Polarization of cell growth in yeast. I. Establishment and maintenance of polarity states. Journal of Cell Science. 2000;**113**(3):365-375

[56] Guo W, Sacher M, Barrowman J, Ferro-Novick S, Novick P. Protein complexes in transport vesicle targeting. Trends in Cell Biology. 2000;**10**(6):251-255. DOI: 10.1016/S0962-8924(00)01754-2

[57] Waters MG, Hughson FM. Membrane tethering and fusion in the secretory and endocytic pathways. Traffic. 2000;**1**:588-597. DOI: 10.1034/j.1600-0854.2000.010802.x

[58] He B, Xi F, Zhang X, Zhang J, Guo W. Exo70 interacts with phospholipids and mediates the targeting of the exocyst to the plasma membrane. The EMBO journal. 2007;**26**(18): 4053-4065. DOI: 10.1038/sj.emboj.7601834

[59] Forsmark A, Rossi G, Wadskog I, Brennwald P, Warringer J, Alder L. Quantitative proteomics of yeast post-Golgi vesicles reveals a discriminating role for Sro7p in protein secretion. Traffic, research support, non-U.S. Gov't. 2011;**12**(6):740-753. DOI: 10.1111/j.1600-0854.2011.01186.x

[60] Coakley G, Maizels RM, Buck AH. Exosomes and other extracellular vesicles: The new communicators in parasite infections. Trends in parasitology. 2015;**31**(10):477-489. DOI: 10.1111/j.1600-0854.2011.01186.x

[61] Oliveira DL, Rizzo J, Joffe LS, Godinho RM, Rodrigues ML. Where do they come from and where do they go: Candidates for regulating extracellular vesicle formation in fungi. International Journal of Molecular Sciences. 2013;**14**(5):9581-9603. DOI: 10.3390/ijms14059581

[62] Oliveira DL, Nakayasu ES, Joffe LS, Guimarães AJ, Sobreira TJ, Nosanchuk JD, Cordero RJ, Frase S, Casadevall A, Almeida IC, Nimrichter L. Characterization of yeast extracellular vesicles: Evidence for the participation of different pathways of cellular traffic in vesicle biogenesis. PLoS One. 2010;**5**:e11113. DOI: 10.1371/journal.pone.0011113

[63] Rayner S, Bruhn S, Vallhov H, Andersson A, Billmyre RB, Scheynius A. Identification of small RNAs in extracellular vesicles from the commensal yeast Malassezia sympodialis. Scientific Reports. 2017;**7**. DOI: 10.1038/srep39742

[64] Ohno S, Ishikawa A, Kuroda M. Roles of exosomes and microvesicles in disease pathogenesis. Advanced Drug Delivery Review. 2013;**65**(3):398-401. DOI: 10.1016/jL.addr.2012.07.019

[65] Oliveira DL, Freire-de-Lima CG, Nosanchuk JD, Casadevall A, Rodrigues ML, Nimrichter L. Extracellular vesicles from Cryptococcus Neoformans modulate macrophage functions. Infection and immunity. 2010;**78**(4):1601-1609. DOI: 10.1371/journal.pone.0021480

[66] Gehrmann U, Qazi KR, Johansson C, Hultenby K, Karlsson M, Lundeberg L, Gabrielsson S, Scheynius A. Nanovesicles from Malassezia sympodialis and host exosomes induce cytokine responses–novel mechanisms for host-microbe interactions in atopic eczema. PLoS One. 2011;**6**(7):e21480. DOI: 10.1371/journal.pone.0021480

[67] Vallejo MC, Matsuo AL, Ganiko L, Medeiros LCS, Miranda K, Silva LS, Freymüller-Haapalainen E, Sinigaglia-Coimbra R, Almeida IC, Puccia R. The pathogenic fungus Paracoccidioides brasiliensis exports extracellular vesicles containing highly immunogenic α-Galactosyl epitopes. Eukaryotic Cell. 2011;**10**(3):343-351. DOI: 10.1128/EC.00227-10

[68] Vargas G, Rocha JD, Oliveira DL, Albuquerque PC, Frases S, Santos SS, Nosanchuk JD, Gomes AM, Medeiros LC, Miranda K, Sobreira TJ. Compositional and immunobiological analyses of extracellular vesicles released by Candida Albicans. Cell. Microbiology. 2015;**17**:389-407. DOI: 10.1111/cmi.12374

[69] Reales-Calderón JA, Vaz C, Monteoliva L, Molero G, Gil C. *Candida albicans* modifies the protein composition and size distribution of THP-1 macrophage-derived extracellular vesicles. Journal of Proteome Research. 2017. DOI: 10.1021/acs.jproteome.6b00605

[70] Baltazar LM, Nakayasu ES, Sobreira TJ, Choi H, Casadevall A, Nimrichter L, Nosanchuk JD. Antibody binding alters the characteristics and contents of extracellular vesicles released by Histoplasma capsulatum. Msphere. 2016;**1**(2):e00085-00015. DOI: 10.1128/mSphere.00085-15

[71] Cordero RJ, Liedke SC, Araújo GRS, Martinez LR, Nimrichter L, Frases S, Peralta JM, Casadevall A, Rodrigues ML, Nosanchuk JD, Guimaraes AJ. Enhanced virulence of Histoplasma capsulatum through transfer and surface incorporation of glycans from Cryptococcus Neoformans during co-infection. Scientific Reports. 2016;**6**. DOI: 10.1038/srep21765

[72] Shaner NC, Steinbach PA, Tsien RY. A guide to choosing fluorescent proteins. Nature Methods. 2005;**2**(12):905-909. DOI: 10.1038/nmeth819

[73] Novick P, Field C, Schekman R. Identification of 23 complementation groups required for post-translational events in the yeast secretory pathway. Cell. 1980;**21**(1):205-215

[74] Burnette WN. "Western blotting": Electrophoretic transfer of proteins from sodium dodecyl sulfate-polyacrylamide gels to unmodified nitrocellulose and radiographic detection with antibody and radioiodinated protein a. Analytical Biochemistry. 1981; **112**(2):195-203. DOI: 10.1016/0003-2697(81)90281-5

[75] Turner TL, Kim E, Hwang C, Zhang G-C, Liu J-J, Jin Y-S. Short communication: Conversion of lactose and whey into lactic acid by engineered yeast. American Dairy Science Association. 2017;**100**:1-5. DOI: 10.3168/jds.2016-11784

[76] Díaz-Hellín P, Naranjo V, Úbeda J, Briones A. Saccharomyces Cerevisiae and metabolic activators: HXT3 gene expression and fructose/glucose discrepancy in sluggish fermentation conditions. World Journal of Microbiology and Biotechnology. 2016;**32**(12):196. DOI: 10.1007/s11274-016-2154-9

[77] Dai C, Xiong F, He R, Zhang W, Ma H. Effects of low-intensity ultrasound on the growth, cell membrane permeability and ethanol tolerance of Saccharomyces Cerevisiae. Ultrasonics Sonochemistry. 2017;**36**:191-197

[78] Romanos MA, Scorer CA, Clare JJ. Foreign gene expression in yeast: A review. Yeast. 1992;**8**(6):423-488. DOI: 10.1002/yea.320080602

[79] Waugh DS. An overview of enzymatic reagents for the removal of affinity tags. Protein Expression and Purification. 2011;**80**(2):283-293. DOI: 10.1016/j.pep.2011.08.005

[80] Kimple ME, Brill AL, Pasker RL. Overview of affinity tags for protein purification. Current Protocols in Protein Science. 2013;**73**(9):9.1-9.9. DOI: 10.1002/0471140864.ps0909s73

[81] Møller TS, Hay J, Saxton MJ, Bunting K, Petersen EI, Kjærulff S, Finnis CJ. Human β-defensin-2 production from S. Cerevisiae using the repressible MET17 promoter. Microbial Cell Factories. 2017;**16**(1):11. DOI: 10.1186/s12934-017-0627-7

[82] Inokuma K, Bamba T, Ishii J, Ito Y, Hasunuma T, Kondo A. Enhanced cell-surface display and secretory production of cellulolytic enzymes with Saccharomyces Cerevisiae

Sed1 signal peptide. Biotechnology and Bioengineering. 2016;**113**(11):2358-2366. DOI: 10.1002/bit.26008

[83] Davison SA, den Haan R, van Zyl WH. Heterologous expression of cellulase genes in natural Saccharomyces Cerevisiae strains. Applied Microbiology and Biotechnology. 2016;**100**(18):8241-8254. DOI: 10.1007/s00253-016-7735-x

[84] Spagnoli G, Bolchi A, Cavazzini D, Pouyanfard S, Müller M, Ottonello S. Secretory production of designed multipeptides displayed on a thermostable bacterial thioredoxin scaffold in Pichia Pastoris. Protein expression and purification. 2017;**129**:150-157. DOI: 10.1016/j.pep.2016.04.012

[85] Chahal S, Wei P, Moua P, Park SPJ, Kwon J, A Patel, Vu AT, Catolico JA, Tsai YF, Shaheen N, Chu TT. Structural characterization of the α-mating factor prepro-peptide for secretion of recombinant proteins in Pichia pastoris. Gene. 2017;**598**:50-62. DOI: 1016/j.gene.2016.10.040

[86] Cimini D, Della Corte K, Finamore R, Andreozzi L, Stellavato A, Pirozzi AV, Ferrara F, Formisano R, De Rosa M, Chino M, Lista L. Production of human pro-relaxin H2 in the yeast Pichia Pastoris. BMC Biotechnology. 2017;**17**(1):4. DOI: 10.1186/s12896-016-0319-0

[87] Jallouli R, Parsiegla G, Carrière F, Gargouri Y, Bezzine S. Efficient heterologous expression of Fusarium solani lipase, FSL2, in Pichia Pastoris, functional characterization of the recombinant enzyme and molecular modeling. International Journal of Biological Macromolecules. 2017;**94**:61-71. DOI: 10.1016/j.ijbiomac.2016.09.030

[88] Abdulrachman D, Thongkred P, Kocharin K, Nakpathom M, Somboon B, Narumol N, Champreda V, Eurwilaichitr L, Suwanto A, Nimchua T, Chantasingh D. Heterologous expression of Aspergillus aculeatus endo-polygalacturonase in Pichia Pastoris by high cell density fermentation and its application in textile scouring. BMC Biotechnology. 2017;**17**(1):15. DOI: 10.1186/s12896-017-0334-9

[89] Boumaiza M, Chahed H, Ezzine A, Jaouan M, Gianoncelli A, Longhi G, Carmona F, Arosio P, Sari MA, Marzouki MN. Recombinant overexpression of camel hepcidin cDNA in Pichia Pastoris: Purification and characterization of the polyHis-tagged peptide HepcD-his. Journal of Molecular Recognition. 2017;**30**(1). DOI: 10.1002/jmr.2561

[90] Vallet-Courbin A, Larivière M, Hocquellet A, Hemadou A, Parimala S-N, Laroche-Traineau J, Santarelli X, Clofent-Sanchez G, Jacobin-Valat MJ, Noubhani A. A recombinant human anti-platelet scFv antibody produced in Pichia Pastoris for atheroma targeting. PLoS One. 2017;**12**(1):e0170305). DOI: 10.1371/journal.pone.0170305

[91] Zhou K, Dong Y, Zheng H, Chen B, Mao R, Zhou L, Wang Y. Expression, fermentation, purification and lyophilisation of recombinant Subtilisin QK in Pichia Pastoris. Process Biochemistry. 2017. DOI: 10.1016/j.procbio.2016.12.028

[92] Dulermo R, Brunel F, Dulermo T, Ledesma-Amaro R, Vion J, Trassaert M, Thomas S, Nicaud JM, Leplat C. Using a vector pool containing variable-strength promoters to

optimize protein production in Yarrowia lipolytica. Microbial Cell Factories. 2017;**16**(1): 31. DOI: 10.1186/s12934-017-0647-3

[93] Spohner SC, Czermak P, Heterologous P. Expression of Aspergillus terreus fructosyltransferase in Kluyveromyces lactis. New Biotechnology. 2016;**33**:473-479. DOI: 10.1016/j.nbt.2016.04.001

[94] Stressler T, Leisibach D, Lutz-Wahl S, Kuhn A, Fischer L. Homologous expression and biochemical characterization of the arylsulfatase from Kluyveromyces lactis and its relevance in milk processing. Applied Microbiology and Biotechnology. 2016;**100**:5401-5414. DOI: 10.1007/s00253-016-7366-2

[95] Madhavan A, Sukumaran RK. Secreted expression of an active human interferon-beta (HuIFNβ) in Kluyveromyces lactis. Engineering in Life Sciences. 2016;**16**(4):379-385. DOI: 10.1002/elsc.201500120

[96] Nielsen J. Production of biopharmaceutical proteins by yeast: Advances through metabolic engineering. Bioengineered. 2013;**4**(4):207-211. DOI: 10.4161/bioe.22856

[97] Hamilton SR, Davidson RC, Sethuraman N, Nett JH, Jiang Y, Rios S, Bobrowicz P, Stadheim TA, Li H, Choi BK, Hopkins D. Humanization of yeast to produce complex terminally sialylated glycoproteins. Science. 2006;**313**(5792):1441-1443. DOI: 10.1126/science.1130256

[98] Solá RJ, Griebenow K. Glycosylation of therapeutic proteins. BioDrugs. 2010;**24**:9-21. DOI: 10.2165/11530550-000000000-00000

[99] Poirier Y, Erard N, Petétot JM-C. Synthesis of polyhydroxyalkanoate in the peroxisome of Saccharomyces Cerevisiae by using intermediates of fatty acid β-oxidation. Applied and environmental microbiology. 2001;**67**(11):5254-5260. DOI: 10.1128/AEM.67.11.5254-5260.2001

[100] Poirier Y, Erard N, Petétot JMC. Synthesis of polyhydroxyalkanoate in the peroxisome of Pichia Pastoris. FEMS Microbiology Letters. 2002;**207**(1):97-102. DOI: 10.1111/j.1574-6968.2002.tb11035.x

[101] Chen L, Zhang J, Chen WN. Engineering the Saccharomyces Cerevisiae β-oxidation pathway to increase medium chain fatty acid production as potential biofuel. PLoS One. 2014;**9**(1):e84853). DOI: 10.1371/journal.pone.0084853

[102] Marchesini S, Erard N, Glumoff T, Hiltunen JK, Poirier Y. Modification of the monomer composition of Polyhydroxyalkanoate synthesized in Saccharomyces Cerevisiae expressing variants of the -oxidation-associated multifunctional enzyme. Applied and environmental microbiology. 2003;**69**(11):6495-6499. DOI: 10.1128/AEM.69.11.6495-6499.2003

[103] Sheng J, Stevens J, Feng X. Pathway compartmentalization in peroxisome of Saccharomyces Cerevisiae to produce versatile medium chain fatty alcohols. Scientific Reports. 2016;**6**. DOI: 10.1038/srep26884

[104] Rutter CD, Rao CV. Production of 1-decanol by metabolically engineered Yarrowia lipolytica. Metabolic engineering. 2016;**38**:139-147. DOI: 10.1016/j.ymben.2016.07.011

[105] Fillet S, Gibert J, Suárez B, Lara A, Ronchel C, Adrio JL. Fatty alcohols production by oleaginous yeast. Journal of industrial microbiology & biotechnology. 2015;**42**(11):1463-1472. DOI: 10.1007/s10295-015-1674-x

[106] Zhou YJ, Buijs NA, Zhu Z, Gómez DO, Boonsombuti A, Siewers V, Nielsen J. Harnessing yeast peroxisomes for biosynthesis of fatty-acid-derived biofuels and chemicals with relieved side-pathway competition. Journal of the American Chemical Society. 2016;**138**(47):15368-15377. DOI: 10.1021/jacs.6b07394

[107] Polach KJ, Neef DW, Fewell JG, Anwer K. Compositions and Methods for Yeast Extracellular Vesicles as Delivery Systems. Google Patents; 2016

Section 4

Molecular Biology of the Yeast

Chapter 6

Nucleosome Positioning and Its Role in Gene Regulation in Yeast

Hongde Liu, Weiheng Ma, Jiahao Xie, Huamei Li,
Kun Luo, Donghui Luo, Lei Liu and Xiao Sun

Additional information is available at the end of the chapter

http://dx.doi.org/10.5772/intechopen.70935

Abstract

Nucleosome, composed of a 147-bp segment of DNA helix wrapped around a histone protein octamer, serves as the basic unit of chromatin. Nucleosome positioning refers to the relative position of DNA double helix with respect to the histone octamer. The positioning has an important role in transcription, DNA replication and other DNA transactions since packing DNA into nucleosomes occludes the binding site of proteins. Moreover, the nucleosomes bear histone modifications thus having a profound effect in regulation. Nucleosome positioning and its roles are extensively studied in model organism yeast. In this chapter, nucleosome organization and its roles in gene regulation are reviewed. Typically, nucleosomes are depleted around transcription start sites (TSSs), resulting in a nucleosome-free region (NFR) that is flanked by two well-positioned H2A.Z-containing nucleosomes. The nucleosomes downstream of the TSS are equally spaced in a nucleosome array. DNA sequences, especially 10–11 bp periodicities of some specific dinucleotides, partly determine the nucleosome positioning. Nucleosome occupancy can be determined with high throughput sequencing techniques. Importantly, nucleosome positions are dynamic in different cell types and different environments. Histones depletions, histones mutations, heat shock and changes in carbon source will profoundly change nucleosome organization. In the yeast cells, upon mutating the histones, the nucleosomes change drastically at promoters and the highly expressed genes, such as ribosome genes, undergo more change. The changes of nucleosomes tightly associate the transcription initiation, elongation and termination. H2A.Z is contained in the +1 and −1 nucleosomes and thus in transcription. Chaperon Chz1 and elongation factor Spt16 function in H2A.Z deposition on chromatin. The chapter covers the basic concept of nucleosomes, nucleosome determinant, the techniques of mapping nucleosomes, nucleosome alteration upon stress and mutation, and Htz1 dynamics on chromatin.

Keywords: nucleosome, 10–11 bp periodicities, nucleosome-free region, MNase-sequencing, histone mutation, H2A.Z

1. Basic conceptions about chromatin and nucleosome

1.1. Chromatin of eukaryotic DNA, nucleosome, nucleosome compositions, and histone

1.1.1. Chromatin

Eukaryotic DNA exists as chromatin structure, which is composed of DNA and proteins in the nucleus (**Figure 1**). The proteins can divide into histone proteins (H1/H5, H2A, H2B, H3, and H4) and non-histone ones. The former acts as core which DNA winds. The histone winding with DNA acts as a ball that forms the basic structure. Non-histone proteins have three main functions: (1) enzyme used in different DNA activities, for example, DNA reparation, duplication, and translation, such as DNA polymerase and DNA ligase; (2) scaffold proteins. They play the role of skeleton; and (3) other motor proteins. All play essential roles in cell structure and regulatory functions that make life possible.

Since the package of DNA must be rapidly accessible so that protein machinery is able to interact with DNA in replication, transcription, DNA repair, and recombination, the chromatin is highly different in different cells and different periods. Chromatin can be divided into euchromatin and heterochromatin. Heterochromatin is characterized by its high compactness and its inhibitory effect on DNA transactions such as gene expression. However, according to Volpe et al. [1], many of them actually can transcribe but are silenced by RNA by RNA-induced transcriptional silencing (RITS). Euchromatin is the chromatin which is not packaged tightly like heterochromatin so it is more accessible. Most of chromatin are euchromatin (92% of the human genome [2]); it contains activating genes and changes its condensation during cell cycle.

1.2. Nucleosome and histones

Nucleosomes are the basic unit of chromatin. The nucleosome consists of 147 bp of DNA wrapped around an octamer of histones, with two copies of each H2A, H2B, H3, and H4, and about 1.65 superhelical turns arranged in a left-handed manner [3] (**Figure 2**). The nucleosome cores are connected by linker DNA, which typically ranges from 10 bp to 90 bp in length, to form a "beads-on-a-string" nucleosomal array with a diameter of 11 nm [4]. At the entry and the exit of the nucleosome, H1 binds the DNA to make the nucleosomes fixate in the space.

The "tails" of these histone proteins stick out, especially H3 and H4, where they can be modified in many ways. Modifications of the tail include methylation, acetylation, phosphorylation, ubiquitination, SUMOylation, citrullination, and ADP-ribosylation. Through these chemical modifications, histone can change its interaction with DNA. Interestingly, many of these modifications have fixed position and function (**Table 1**).

All histones have its variants (**Figure 3**) [5], and they have different biological function compared to canonical histones. Exchanging with canonical histones dynamically also plays an important role in regulation of gene expression.

http://dx.doi.org/10.5772/intechopen.70935

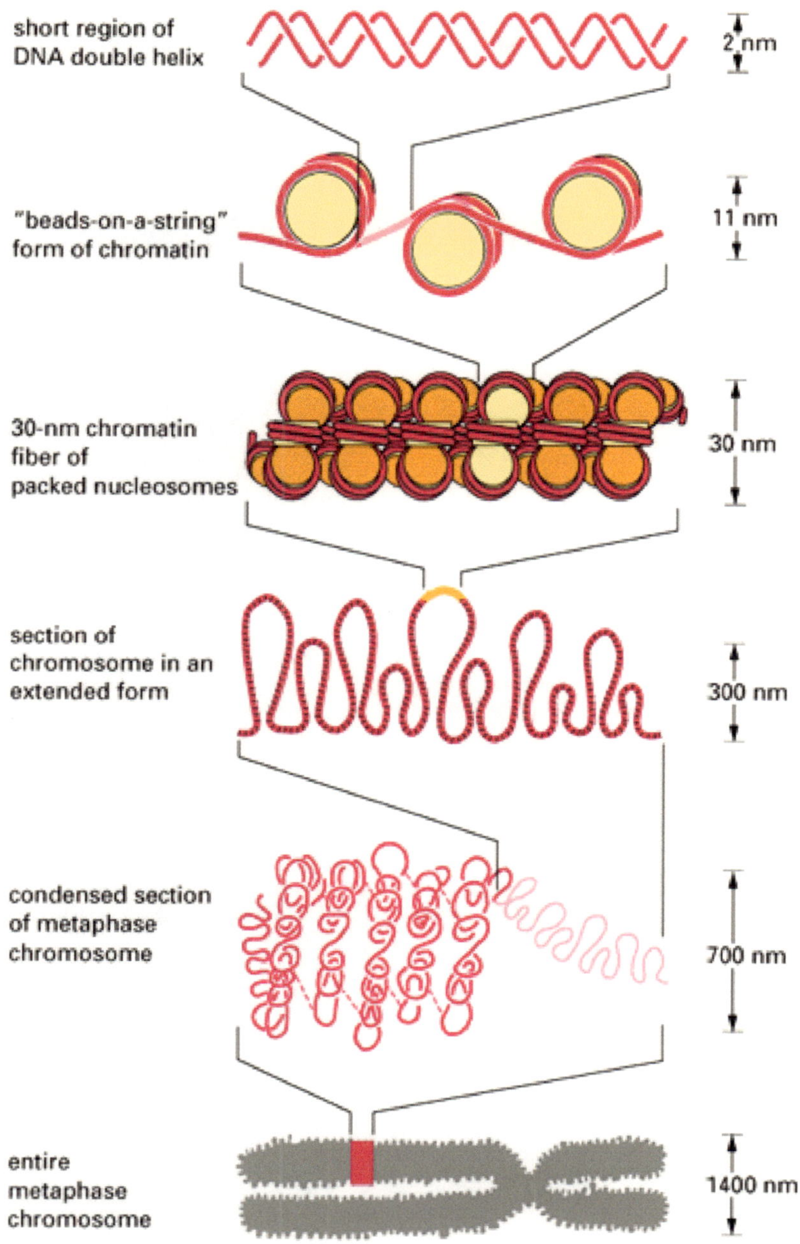

Figure 1. Chromatin structure (http://csma31.csm.jmu.edu/chemistry/faculty/mohler/chromatin.htm).

1.3. Research history of nucleosomes, especially in yeast

Clark and Felsenfeld first used staphylococcal nuclease to digest chromatin in 1971 and found that some regions were sensitive to nuclease while some were insensitive; insensitive regions were homogeneous, suggesting it contains subunits. Then Hewish and Burgoyun Researchers

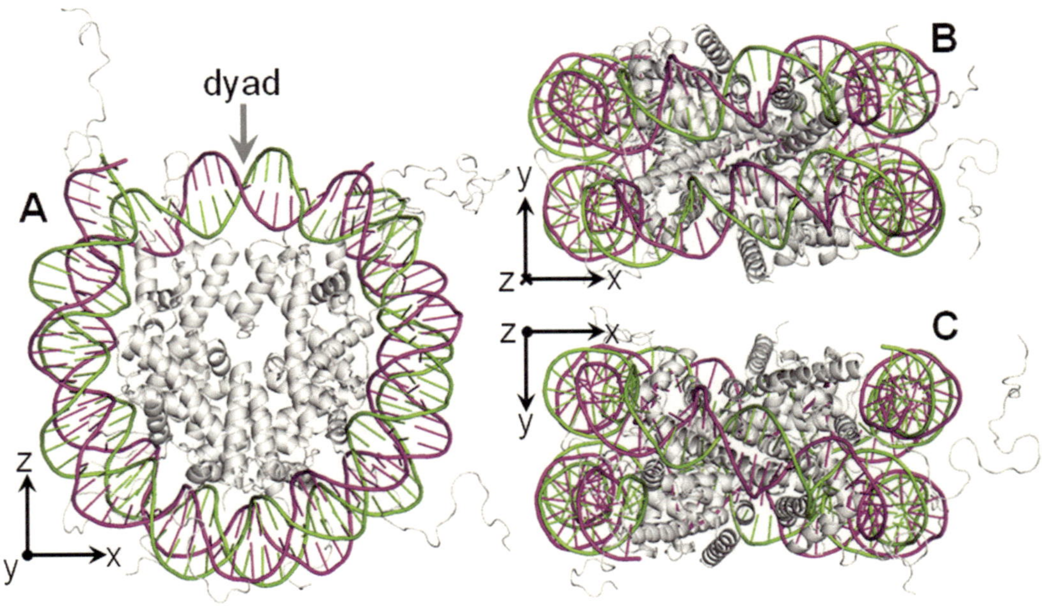

Figure 2. These views of NCP147 (147-bp nucleosome core particle), at Å resolution, show the two strands of the double-helix in purple and green, with the protein core in gray. (A) The curvature of DNA around the histone core, with the dyad at the top, center; (B) a rotation of the particle, showing the adjacent segments of DNA, opposite the dyad; and (C) a rotation in the opposite direction, showing the DNA crossing over the dyad. As indicated by the coordinate system axes, in (A) the y-axis is pointing out of the page, in (B) the z-axis is pointing into the page, and in (C) the z-axis is pointing out of the page [6].

Type of modification	Histone								
	H3K4	H3K9	H3K14	H3K27	H3K79	H3K36	H4K20	H2BK5	H2BK20
Mono-Me	A	A		A	A		A	A	
Di-Me		R		R	A				
Tri-Me	A	R		R	A & R	A		R	
Ac		A	A	A					A

Note: Me, methylation; Ac, acetylation; A, activation; R, repression.

Table 1. Different modifications in transcription regulation (from) [7–10].

in previous study digested the nuclei with endogenous nuclease and isolated DNA from the nucleus. As a result, a series of DNA fragments were found, which corresponded to a basic unit of about 200 bp, indicating that histones bind to DNA in a regular manner which results in only certain restricted regions are sensitive to nuclease.

Kornberg and Thomas then digested the chromatin with a small cellulase in 1974 and centrifuged it to obtain monomers, dimers, trimers, and tetramers. Using electron microscopy, the monomer was observed as a 10 nm body, and the dimer was two associated bodies. The same

http://dx.doi.org/10.5772/intechopen.70935

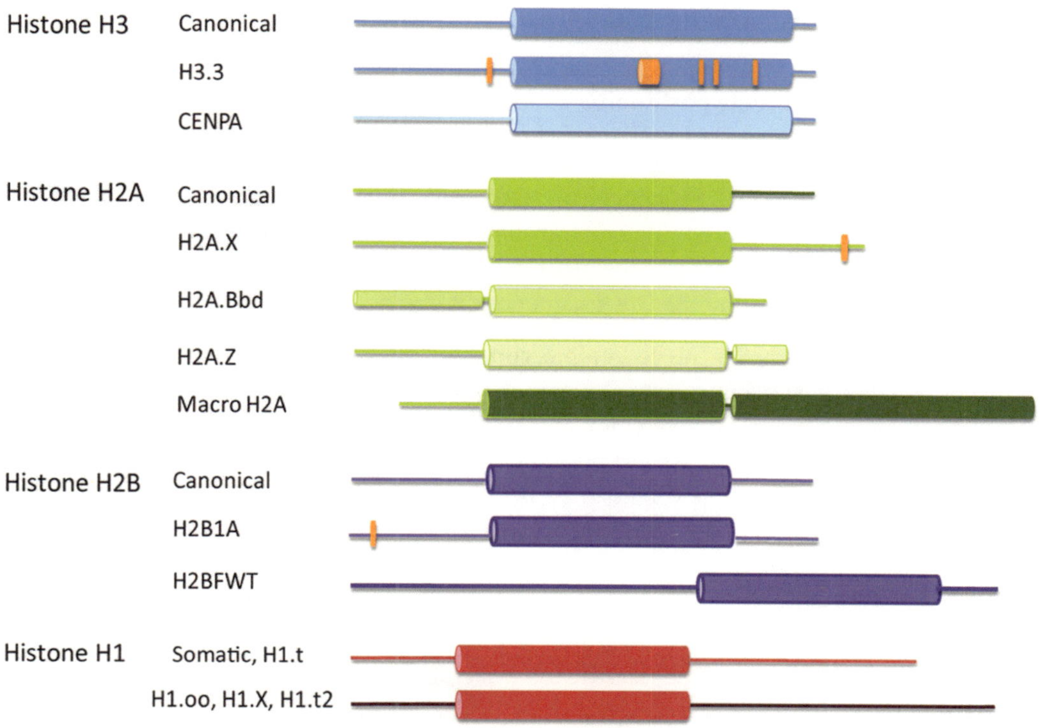

Figure 3. All of the histone variants contain a highly conserved histone fold domain and vary mainly in their C- and N-terminal sequences. The above shown is a schematic comparing histone variant sequences. Boxes represent the histone fold domain and orange lines represent site-specific sequence variations. Histones that are in different shades of the same color are from the same histone family but have large differences in sequence [5].

trimer and tetramer consisted of three bodies and four bodies, respectively, indicating that the structure consisting of 200 bp DNA was "rope beads" units, which are called nucleosomes.

Through all kinds of experiments, it was found that the structure of the nucleosome core is relatively invariant from yeast to metazoans [11, 12] containing a 147 bp DNA wrapped around a histone protein octamer. In 2005, Yuan et al. developed a tiled microarray approach to identify at high resolution the translational positions of 2278 nucleosomes over 482 kb of *Saccharomyces cerevisiae* DNA, including almost all of chromosome III and 223 additional regulatory regions [13]. However, the study of the location of nucleosomes is quite time-consuming and costly if using experiments alone, so the researchers began to build nucleosome positioning prediction model based on the existing experimental data [14]. In yeast genome, Segal et al. found that DNA sequence contains ~10-bp period pattern of AA-TT-TA/GC dinucleotides [15]. Nucleosomal DNA sharp bending occurs at every DNA helical repeat (~10 bp), when the major groove of the DNA faces inward toward the histone octamer, and again ~5 bp away, with opposite direction, when the major groove faces outward. The property of the ~10-bp periodicity is called "a genome code" for nucleosomes. Since that, many nucleosome prediction models were developed.

2. Nucleosome positioning and its determinant

2.1. Concepts of nucleosome positioning and nucleosome occupancy

The term "nucleosome positioning" is used to indicate where nucleosomes are located with respect to the genomic DNA sequence [16]. Generally, nucleosome positioning can divide into two parts: rotational positioning and translational positioning. The first one is to describe the side of the DNA helix that faces the histones and the next one is to determine the nucleosome midpoint with regard to the DNA sequence.

By doing statistical analysis, "nucleosome occupancy" tries to identify the possibilities of a base pair whether it is in a nucleosome region [16]. It is possible to calculate average nucleosome positioning levels on a given region of DNA in a population of cells. In ideal conditions, nucleosome is "shaking" in the perfect position. By counting the time of sequenced reads that are overlapped by nucleosome center in a ~147 bp window, it gives the most conservative locus which means that it is most possible to have a nucleosome there (**Figure 4**).

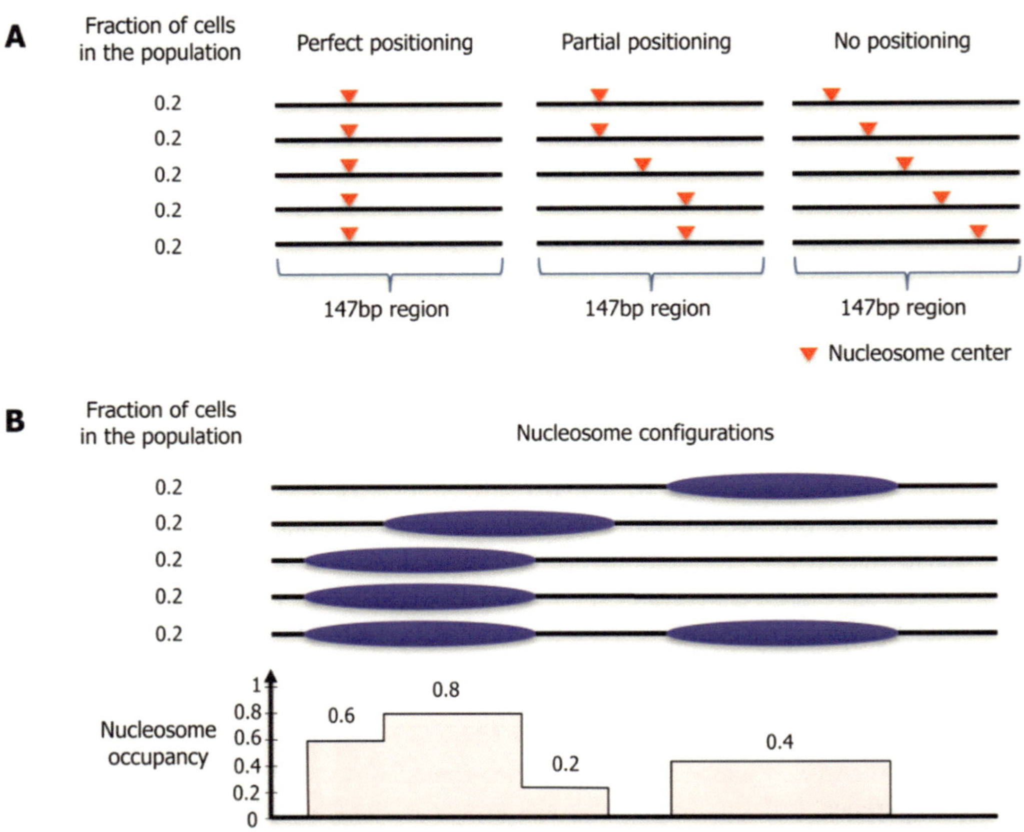

Figure 4. Illustration of the concepts of nucleosome positioning and nucleosome occupancy. (A) We use fraction of cells from the population in which that basepair is in the middle of a 157bp nucleosome representing the nucleosome positioning along every basepair in the genome. The left figure exhibit perfect-positioning region, where the nucleosome center is located at the same basepair all over population cells; the other two showed partial-positioning and no-positioning region. (B) [16].

http://dx.doi.org/10.5772/intechopen.70935

2.2. The association between nucleosome positioning and 10–11 bp periodicities in DNA sequence in yeast

Early in 1990, the 10–11 bp periodicities were reported [17]. In addition to 3-bp periodicity, which is due to the fact that three consecutive bases encode one type of amino acids, the genomic DNA exhibits 10–11 bp periodicities. The 10–11 bp periodicities in complete genomes reflect protein structure and DNA folding [17].

In alpha helices structure, the hydrophobic amino acids (aa) occur with a ~3.5 aa, and all five hydrophobic amino acids L, I, V, F, and M have a base T (thymine) at middle position of their codons. This leads ~3.5 × 3 = ~10.5 bp periodicity in protein coding DNA sequences, called protein-induced periodicity.

On the other hand, the 10–11 bp periodicities have an intimate association with nucleosome positioning. To sharply bent and tightly wrapped around a histone protein octamer, DNA sequence has intrinsic bias. The position of certain dinucleotides, such as AA, TA, and TT in minor grooves facing toward (every 10 bp) and GG in minor grooves facing away from the histone octamer favors these (**Figure 5**) distortions [15]. Moreover, when digesting DNA using DNase I (Deoxyribonuclease I), it was observed that the cleavage pattern in nucleosome position shows a ~10.3 bp period, which is equal to a minor groove. For the naked DNA, which is entirely devoid of nucleosomes, the oscillatory pattern in cleavage profile was disappeared in digesting [18]. All of these strongly suggested the role of the 10–11 bp periodicities of the specific dinucleotides in positioning nucleosomes. Based on the features of DNA sequences, many models are developed to predict nucleosomes (**Table 2**).

We further found that within frequency domains, weakly bound dinucleotides (AA, AT, and the combinations AA-TT-TA and AA-TT-TA-AT) present doublet peaks in a periodicity range

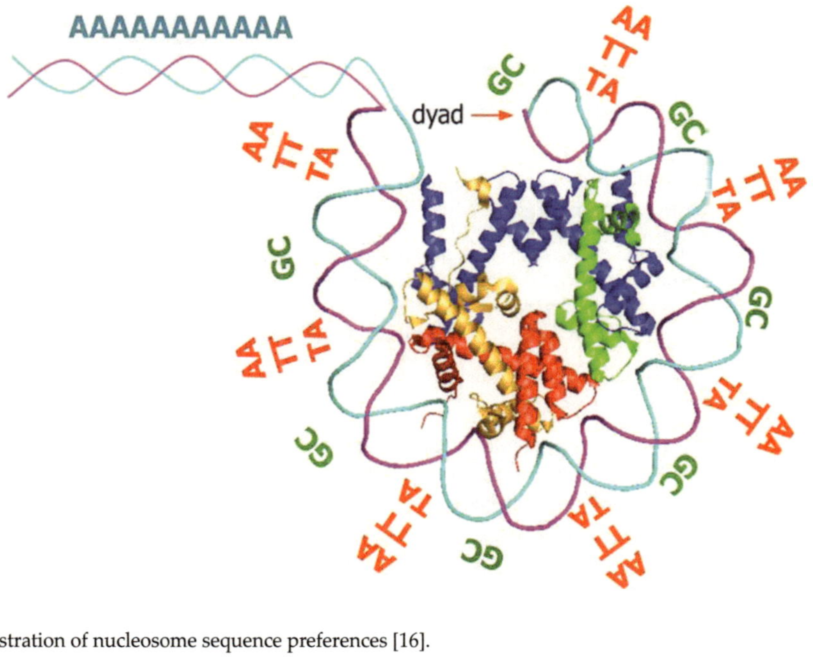

Figure 5. Illustration of nucleosome sequence preferences [16].

Software	Description	References
Segal Lab: Online Nucleosomes Prediction	This tool allows you to submit a genomic sequence and to receive a prediction of the nucleosomes positions on it, based on the nucleosome-DNA interaction model	[15, 21, 22]
iNuc-PseKNC	A predictor for predicting nucleosome positioning in *Homo sapiens*, *Caenorhabditis elegans*, and *Drosophila melanogaster* genomes	[23]
NuCMap	NuCMap is based on chemical modification of engineered histones. The tool reveals novel aspects of the in vivo nucleosome organization that are linked to transcription factor (TF) binding, RNA polymerase pausing, and the higher order structure of the chromatin fiber	[24]
NuPoP	Predicts nucleosome position by explicitly modeling the linker DNA length. NuPoP is based on a duration hidden Markov model (HMM)	[25]
Epidaurus	Epidaurus is a bioinformatics tool used to effectively reveal inter-dataset relevance and differences through data aggregation, integration, and visualization	[26]
Multi-Layer Model	Analyses nucleosome position data obtained with microarray-based approach. MLM is a classifier to distinguish between several kinds of patterns	[27]
NucEnerGen	Predicts nucleosome energetics by using high throughput sequencing. It establishes that nucleosome occupancies can be explained by systematic differences in mono- and dinucleotide content between nucleosomal and linker DNA sequences	[28]
nuMap	Implements the YR and W/S schemes to predict nucleosome positioning at high resolution. This methodology is based on the sequence-dependent anisotropic bending	[29, 30]
NPRD	Compiles the available experimental data on locations and characteristics of nucleosome formation sites (NFSs). The object of the database is a single NFS described in an individual entry	[31]
AWNFR	An algorithm based on down-sampling operation and footprint in wavelet	[32]
ICM	Allows users to assess nucleosome stability and fold sequences of DNA into putative chromatin templates. It uses an elastic model to place nucleosomes	[33]
SymCurv	The tool is able to capture sequence constraints, which are related to structure in genomic regions	
FineStr	Allows users to upload genomic sequences in FASTA format and to perform a single-base-resolution nucleosome mapping on them	[34]
iNuc-PhysChem	Identifies nucleosomal sequences by incorporating physicochemical properties into a 1788-dimensional feature vector. iNuc-PhysChem was able to identify nucleosome positioning for an independent DNA segment extracted from the *Saccharomyces cerevisiae* genome	[35]
TemplateFilter	High-resolution nucleosome mapping reveals transcription-dependent promoter packaging	[36]
DANPOS	A comprehensive bioinformatics pipeline explicitly designed for dynamic nucleosome analysis at single-nucleotide resolution. DANPOS is also robust in defining functional dynamic nucleosomes	[37]
BINOCh	A package that allows biologists to carry out an analysis of nucleosome occupancy data to discover stimulus-induced transcription factor binding	[38]

Software	Description	References
PING	A package for nucleosome positioning using MNase-seq data or MNase- or sonicated ChIP-seq data. PING uses a model-based approach, which enables nucleosome predictions even in the presence of low read counts	[39]
ChIPseqR	A package based on an algorithm for the analysis of nucleosome positioning and histone modification ChIP-seq experiments	[40]
NUCwave	A bioinformatic tool that generates nucleosome occupation maps from chromatin digestion with micrococcal nuclease (MNase-seq), chemical cleavage (CC-seq), chromatin inmunoprecipitation (ChIP-seq) and fragmentation by sonication	[41]
NucPosSimulator	A simulation tool to identify positions of nucleosomes from next generation sequencing data	[42]
NucHunter	Inferring nucleosome positions with their histone mark annotation from ChIP data	[43]
DiNuP	A systematic approach to identify regions of differential nucleosome positioning	[44]
NucTools	Allows calculations of nucleosome occupancy profiles averaged over several replicates, comparisons of nucleosome occupancy landscapes between different experimental conditions, and the estimation of the changes of integral chromatin properties. NucTools facilitates the annotation of nucleosome occupancy with other chromatin features like binding of transcription factors (TF) or architectural proteins, and epigenetic marks like histone modifications or DNA methylation	[45]
Dimnp	Identifies differential nucleosome regions (DNRs) in multiple samples. Dimnp is able to identify all the DNRs that are identified by two-sample method Danpos. It shows a good capacity (area under the curve >0.87) compared with the manually identified DNRs	[46]
ArchAlign	ArchAlign identifies shared chromatin structural patterns from high-resolution chromatin structural datasets derived from next-generation sequencing or tiled microarray approaches for user defined regions of interest	[47]
SANEFALCON	A tool developed to calculate the fetal fraction for noninvasive prenatal testing based on genome-wide nucleosome profiles, based on single end sequencing of cell-free DNA	[48]
NucDe	An R package mapping nucleosome-linker boundaries from both MNase-Chip and MNase-seq data using a non-homogeneous hidden-state model based on first-order differences of experimental data along genomic coordinates	
Nu-OSCAR	A program that can be used to identify binding sites of known transcription factors	
NSeq	A multithreaded Java application for finding positioned nucleosomes from sequencing data	[49]
ArchTEx	The extension of mapped sequence tags is a common step in the analysis of single-end next-generation sequencing (NGS) data from protein localization and chromatin studies. ArchTEx identifies the optimal extension of sequence tags based on the maximum correlation between forward and reverse tags and extracts and visualizes sites of interest using the predicted extension	[50]

Software	Description	References
PuFFIN	Builds genome-wide nucleosome maps specifically designed to take advantage of paired-end reads. This method can accurately determine a genome-wide set of nonoverlapping nucleosomes without any user-defined parameters	[51]
NPS	A python software package that can identify nucleosome positions given histone-modification ChIP-seq or nucleosome sequencing at the nucleosome level	[52]

Table 2. Nucleosome prediction models and nucleosome sequenced data-processed models.

of 10–11 bp, and strongly bound dinucleotides present a single peak [19]. A time-frequency analysis, based on wavelet transformation, indicated that weakly bound dinucleotides of nucleosomal DNA sequences were spaced smaller (~10.3 bp) at the two ends, with larger (~11.1 bp) spacing in the middle section. The finding was supported by DNA curvature and was prevalent in all core DNA sequences.

We assessed the roles of the 10–11 bp periodicities for different kinds of dinucleotides [20]. Near the transcription start site, the signals reveal a similar feature that the nucleosome organization exhibits (**Figure 6**). But, it seems that the species do not share the same dinucleotides patterns. Furthermore, the dinucleotides patterns are dominant at the specific region of genome, indicating their diverse roles in forming and organizing nucleosomes.

2.3. Nucleosome prediction models for yeast

In **Table 2**, the models for both nucleosome prediction and nucleosome sequencing data processing are listed.

2.4. The chromatin remodeling complex and its roles in altering nucleosomes

Chromatin remodeling complex helps cell to establish the access of genomic DNA for transcription factors. The complexes have two major groups, namely covalent histone-modifying complexes and ATP-dependent chromatin remodeling complexes [53]. They work in a different way.

ATP-dependent chromatin-remodeling enzymes are helicase which use ATP's energy to reposition (slide, twist or loop) nucleosomes along the DNA, expel histones away from DNA or facilitate exchange of histone variants, and thus creating nucleosome-free regions of DNA for gene activation [54]. All known ATP-dependent chromatin complex can be organized into SWI/SNF, ISWI, CHD, and INO80 families. Each family of ATPase has distinct remodeling activities, including incremental nucleosome sliding on DNA in *cis*; the creation of DNA loops on the surface of the nucleosome; eviction of histone H2A/H2B dimers; eviction of the histone octamer; or the exchange of histone octamer subunits within the nucleosome to change its composition [55].

Covalent histone-modifying complexes modify the histone including acetylation, methylation, and phosphorylation which can change the interaction between histone and DNA; for

http://dx.doi.org/10.5772/intechopen.70935

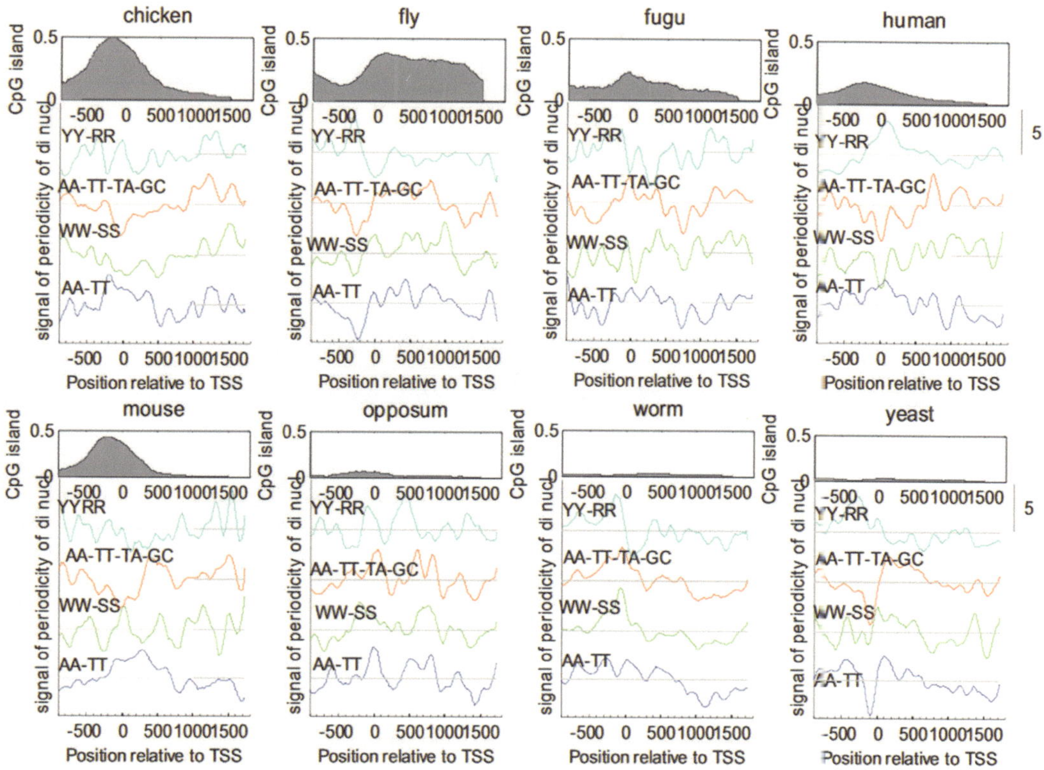

Figure 6. The 10–11 bp periodicities signals of the dinucleotides patterns around TSSs of eight species (human, mouse, chicken, worm, fly, fugu, lancelet, and yeast) [20].

example, methylation of specific lysine residues in H3 and H4 causes further condensation of DNA around histones, making it hard to bind transcription factor or other proteins.

2.5. The statistical model for nucleosomes distribution

A typical nucleosome distribution around TSS is shown in **Figure 7** [56]. Nucleosomes are depleted around TSSs, resulting in a nucleosome-free region (NFR) that is flanked by two well-positioned nucleosomes whereas the nucleosomes downstream of the TSS are equally spaced in a nucleosome array. Of all nucleosomes around the gene, the +1 nucleosome often contains histone variants (H2A.Z and H3.3) and modification by acetyltransferases and methyltransferases. These may help to the nucleosome eviction when transcription is needed. The +2 nucleosome follows the +1 nucleosome immediately and shares the some properties but contains less H2A.Z and less methylation and acetylation. In a barrier model for nucleosome organization, the nucleosome distribution is largely a consequence of statistical packing principles. The genomic sequence specifies the location of the −1 and +1 nucleosomes. The +1 nucleosome forms a barrier against which nucleosomes are packed, resulting in uniform positioning, which decays at farther distances from the barrier [57].

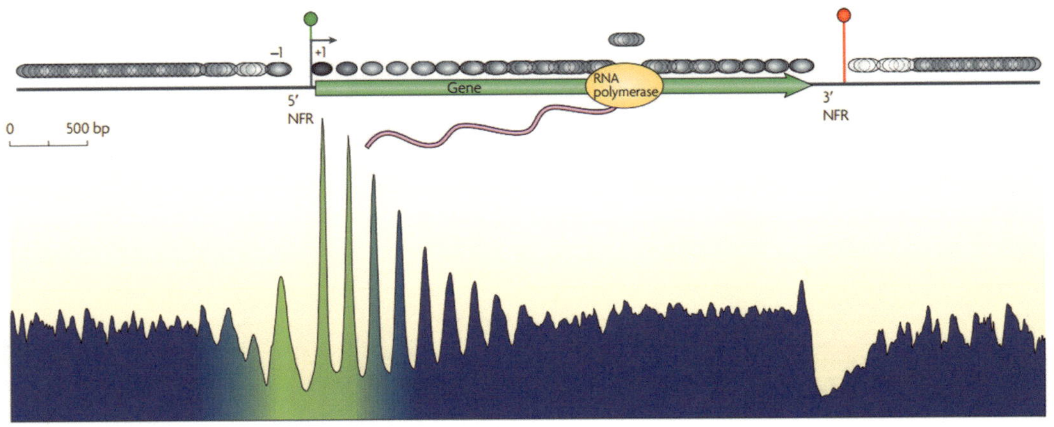

Figure 7. The consensus distribution of nucleosomes (gray ovals) around all yeast genes is shown, aligned by the beginning and end of every gene. The resulting two plots were fused in the genic region. The peaks and valleys represent similar positioning relative to the transcription start site (TSS). The arrow under the green circle near the 5′nucleosome-free region (NFR) represents the TSS. The green-blue shading in the plot represents the transitions observed in nucleosome composition and phasing (green represents high H2A.Z levels, acetylation, H3K4 methylation and phasing, whereas blue represents low levels of these modifications). The red circle indicates transcriptional termination within the 3′ NFR. Figure is reproduced from REF(2008) Cold Spring Harbor Laboratory Press [56].

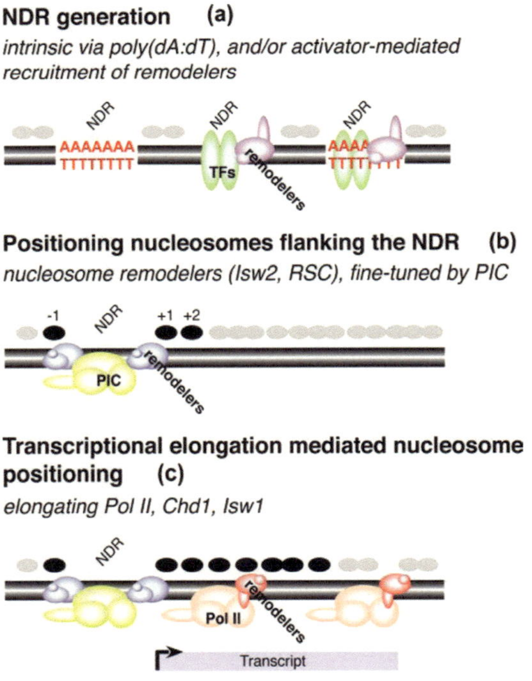

Figure 8. Determinants of nucleosome positioning. (a) Nucleosome-depleted regions (NDRs) are generated either by poly (dA:dT) tracts and/or by transcription factors and their recruited nucleosome remodeling complexes. Gray circles indicate nucleosomes. (b) Nucleosomes located at highly preferred positions (black circles) flanking the NDR are generated by nucleosome-remodeling complexes (for example, Isw2 and RSC, likely in a transcription-independent manner), and fine-tuned by the Pol II preinitiation complex (PIC) and associated factors. (c) Positioning of the more downstream nucleosomes depends on transcriptional elongation, and the recruitment of nucleosome-remodeling activities (for example, Chd1 and Isw1) and histone chaperones by the elongating Pol II machinery [16].

2.6. The nucleosome determinant [16]

A variety of factors determine the location of nucleosomes including DNA sequence, nucleosome remodelers, transcription factors (TFs), and elongating Pol II (**Figure 8**). Each of these components has different contribution in nucleosome positioning. Interestingly, these components can affect each other thus resulting in different positioning pattern in a more complex way. The DNA sequence is critical for rotational positioning along the DNA helix, and it is also an important determinant for nucleosome occupancy. In particular, poly (dA:dT) and poly (dG:dC) tracts are intrinsically inhibitory to nucleosome formation, whereas non-homopolymeric GC-rich regions favor nucleosome formation.

3. The experiment methods of determining nucleosome occupancy and the bioinformatics analysis for the data

3.1. The techniques of determining nucleosomes positions

3.1.1. Mnase-seq

Micrococcal nuclease (MNase), one kind of glycolprotein of *Staphylococcus aureus*, has capacity of digesting the naked DNA. MNase, firstly, induces single-strand breaks, and then cleaves the complementary strand near the first break [58, 59]. Nucleosomal DNA is protected by wrapping on histone octamer in digesting with MNase, thus being remained as DNA fragments after the digestion. Taking this advantage, a high throughput sequencing technique MNase-seq is developed to probe nucleosome positions in a genome-wide manner. MNase cleavage favors AT-rich region in limiting enzyme concentrations.

3.1.2. Dnase-seq

DNase I, one kind of endonuclease, can cut the chromatin-accessible DNA, namely DNase I hypersensitive sites (DHSs), and thus is used in mapping opening chromatin regions (**Figure 9**) [60]. The opening chromatin region is mainly the regulatory sites in gene transcription. Thus, the opening region may alter in different cells types. This can be reflected in DHSs. The change of DHSs often associates one or more nucleosomes loss or formation [60].

DNase-seq means the DNase I digestion followed by DNA sequencing [60]. DNase-seq has been widely used in probing cell-specific chromatin accessibility. The rotational localization of individual nucleosomes is based on the inherent preference of DNA enzyme I cleavage of DNA at about 10 bp per nucleosome [61]. By coupling bioinformatics analysis, DNase-seq can be used in studying TF occupancy at nucleotide resolution in a qualitative and quantitative manner [62]. In DNase-seq, many cells and many sample preparations and enzyme titration steps are required [63].

3.1.3. ATAC-seq

ATAC-seq is an assay for transposase-accessible chromatin with high throughput sequencing [64]. The technique is based on Tn5 transposase's "cutting and pasting" function to probe

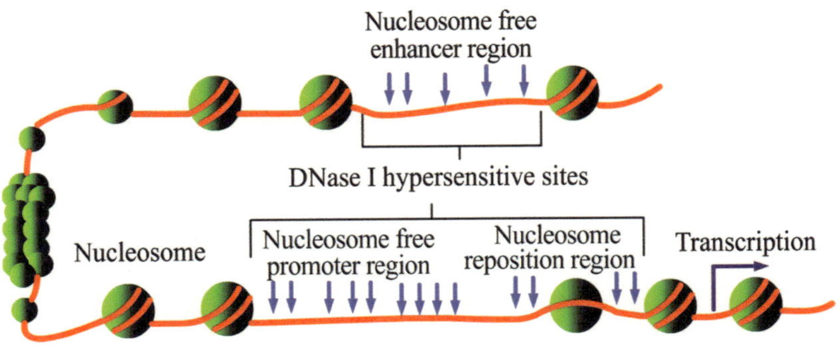

Figure 9. DNase I hypersensitive sites within chromatin [60].

the active regulatory regions [65]. ATAC-seq only needs a small number of cells, ~500–50,000 unfixed nuclei. Moreover, its procedure only involves two steps. Therefore, it is able to study multiple aspects of chromatin architecture simultaneously at high resolution, including nucleosomes, chromatin accessibility [64].

3.1.4. ChIP-seq

Chromatin immunoprecipitation followed by sequencing (ChIP-seq) sequences the interest DNA fragments that are separated and collected from the immunoprecipitation [66]. The main area of ChIP-seq is in precisely mapping for transcription factor-binding sites (TFBSs). **Figure 10** shows a general procedure of a ChIP experiment [66]. This procedure includes the DNA-protein crosslinking with formaldehyde, sonication, immunoprecipitation, reversed crosslinking, and sequencing [66]. Using antibody of the histones, such as histone H3, ChIP-seq is immediately able to determine nucleosome positions.

3.1.5. Other techniques

In addition to the techniques mentioned above, there are other techniques often used, such as Formaldehyde-assisted isolation of regulatory elements (FAIRE-seq) and ChIP-exo. FAIRE-seq is based on the differences in crosslinking efficiencies between DNA and nucleosomes or sequence-specific DNA-binding proteins. Sequencing provides information for regions of DNA that are not occupied by histones [67]. ChIP-exo employs the use of exonucleases to degrade strands of the protein-bound DNA in the 5′–3′ direction to within a small number of nucleotides of the protein binding site [68]. The nucleotides of the exonuclease-treated ends are determined using DNA sequencing.

3.2. Procedures of dealing with the nucleosome DNA sequenced dataset

At the present, nucleosome sequencing dataset are mainly from MNase-seq. In some studies, dataset from ATAC-seq, DNase-seq, and ChIP-seq are used to infer nucleosome positions. A general analysis workflow includes data quality control, mapping, making nucleosome profile, determining nucleosome position, comparing between cell types, and associating with other omics-data (expression data) to find biological meanings.

http://dx.doi.org/10.5772/intechopen.70935

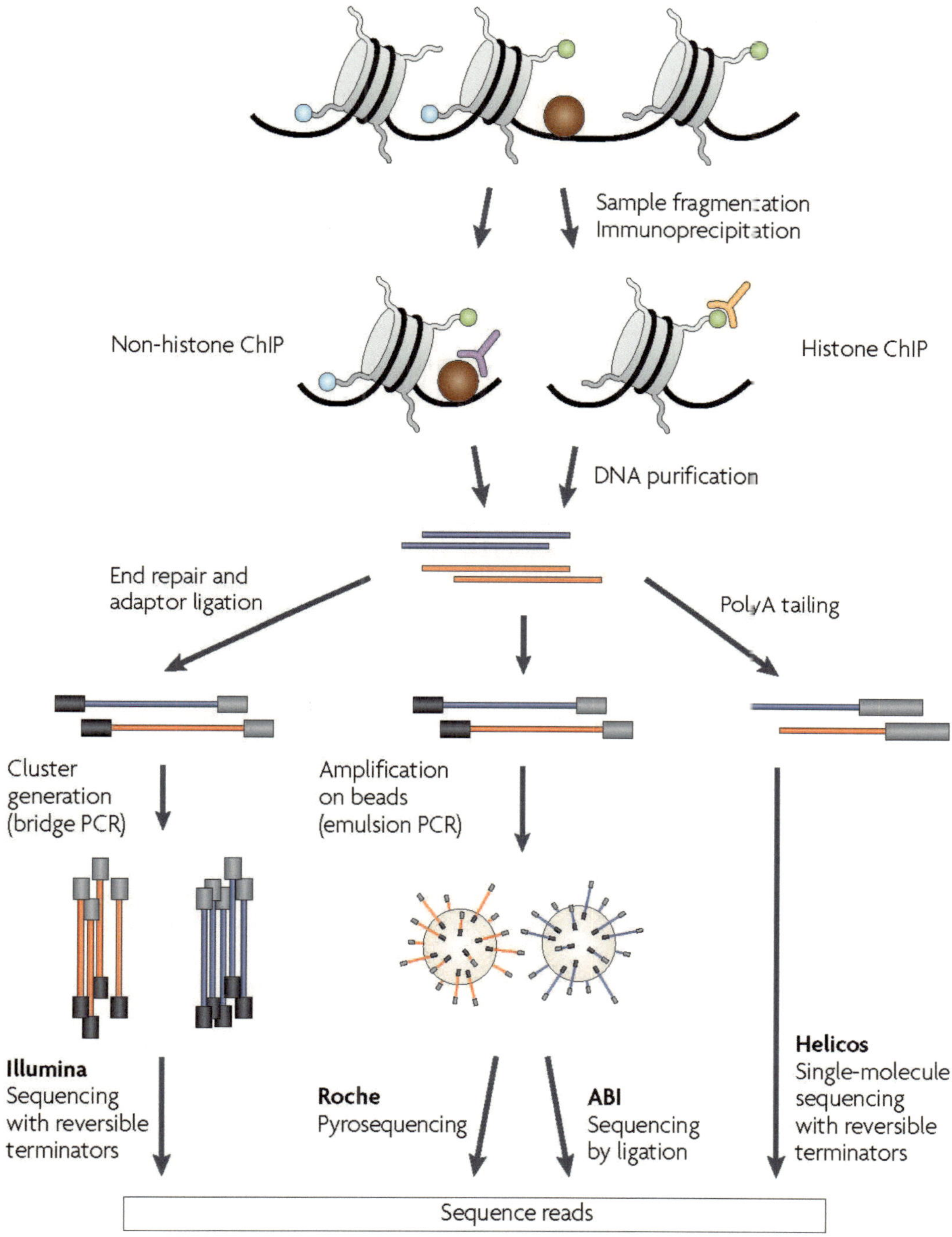

Figure 10. Overview of a ChIP-seq experiment [66] (https://www.nature.com/nrg/journal/v10/n10/full/nrg2641.html).

3.2.1. Data management and genome alignment

Sequencing quality control (QC) is to check the reads quality (fraction of mapped reads) and depth of coverage. Tools BWA and Bowtie are widely used in reads alignments. During the alignment process, multiple-mapping reads and duplication reads are often filtered so as to remove overrepresented regions of the genome due to technical bias [60]. Reads filtering can be performed with SAMtools or Picard tools.

3.2.2. *Data visualization*

Data visualization helps to observe the reads distribution at specific locus. The Integrative Genomics Viewer (IGV) [69], which is developed by the University of California Santa Cruz (UCSC), is one of the most powerful tools to visualize. In IGV, the multiple types of annotation data are integrated, including gene information, epigenetic and expression data, single-nucleotide polymorphisms (SNPs), repeat elements and functional information from the ENCODE, and other research projects. IGV accepts many types of data formation including BED, BedGraph, GFF, WIG, and BAM files, which allow to compare with publicly data.

3.2.3. *Identification of enriched regions*

With respect to nucleosomes sequencing data, there are two basic tasks in analysis. One is to calculate the nucleosome profile (reads coverage) both along the genomic coordinate and near the regulatory sites (for instance the TSSs). This helps to directly check the quality of MNase digestion and DNA sequencing. The other task is to infer the precise nucleosomes positions (dyad position) using the nucleosome profile so as to identify the nucleosome alteration among different cell types.

3.2.3.1. *MNase-seq data*

For single-end MNase-seq data, one method to make nucleosome profile is as follows [70]. First, the length of each read was extended 73 bp in the 3′ direction, and the Watson-strand reads and Crick-strand reads were oppositely shifted 73 bp. The absolute nucleosome occupancy value of each genomic site was expressed as the number of reads covering the genomic sites. Second, nucleosome occupancy was scaled by dividing the occupancy value by the average nucleosome occupancy of the whole genome; i.e., the nucleosome occupancy was expressed as the fold change of the absolute occupancy relative to the average occupancy. Reads can also be shifted 73 bp toward the 3′ direction, which represented the midpoint [60].

With paired-end sequencing, it is assumed that the nucleosome midpoint is consistent with the midpoint of the forward and reverse reads. Unless the reads are from the on type cell (single cell), nucleosome positions actually represent the average positions in cell population. Therefore, the overlapping reads have to be clustered over genomic regions [60].

Calling nucleosomes actually is to find the peak positions along the nucleosome profile. DANPOS is one tool that can identify nucleosome positions [37]. Also, it allows us to detect three categories of nucleosome dynamics, such as position shift, fuzziness change, and occupancy change, using a uniform statistical framework using MNase-seq datasets. Other tools can be found in **Table 2**.

3.2.3.2. *DNase-seq data, ATAC-seq, and ChIP-seq*

From the DNase/ATAC/ChIP-seq datasets, nucleosome position cannot be directly inferred, but they provide information about the opening chromatin and protein-binding regions, which associate nucleosome depletion. Therefore, for these datasets, peaking calling is one central task. MACS identifies genome-wide locations of transcription/chromatin factor binding or histone modification, including removing redundant reads, adjusting read position, calculating peak enrichment, and estimating the empirical false discovery rate (FDR) (http://liulab.dfci.harvard.edu/MACS/index.html) [71]. Based on the position-adjusted reads, MACS slides a window of size 2d across the genome to identify regions that are significantly enriched relative to the genome background. The P-value is derived from Poisson distribution. When a control sample is available, MACS can also estimate an empirical false discovery rate (FDR) for every peak by exchanging the ChIP-seq and control samples and identifying peaks in the control sample using the same set of parameters used for the ChIP-seq sample.

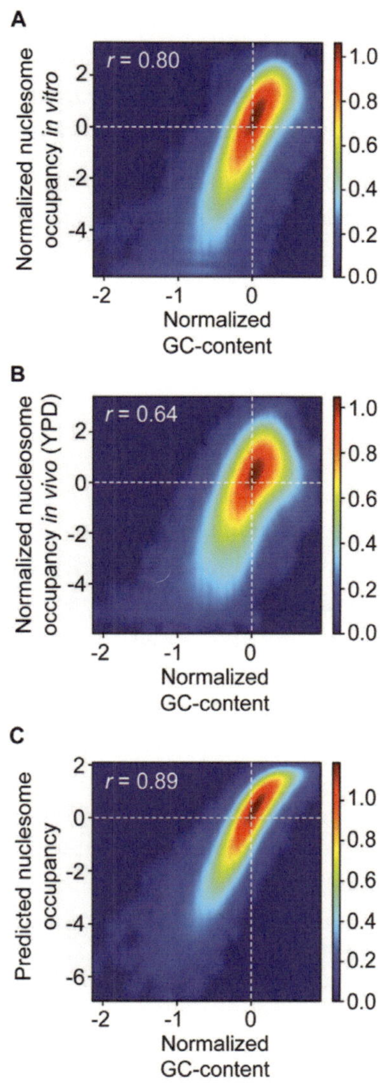

Figure 11. The GC content in 147 base pair windows is strongly correlated to nucleosome occupancy. Density plot comparison between the normalized centered GC content in 147 base pair windows (x axis) and (A) the in vitro reconstituted, (B) the in vivo (YPD). GC-content is normalized as the log2 GC-content within 147 base pairs [75].

3.3. GC-content and cutting bias

GC content bias means the variability between the GC content in a region and the count of fragments/reads mapped to it. The bias can dominate the signal of interest for analyses and leads to false positive. More seriously, the bias tends to be different among the samples; thus, there is no general method to remove it [72]. Two facts associate the variability. One is GC content which is heterogeneous among the genome. In yeast, the open reading frames (ORFs) with similar GC contents at silent codon positions are significantly clustered on chromosomes [73]. Moreover, GC content varies along the genome and is often correlated with functionality. The other is MNase that has a cutting bias. Kinetic analysis indicates that the rate of cleavage is 30 times greater at the 5′ side of A or T than at G or C [74].

Most current correction methods follow a common path. Both fragment counts and GC counts are binned to a bin-size of choice [72]. Then, the conditional mean fragment count per GC value is modeled by assuming smoothness. At last, a predicted count is estimated for each bin based on the bin's GC. The predictions represent one normalization for the original signal [72].

Another aspect is that GC-content is predictive for nucleosome position both in vivo and in vitro (**Figure 11**) [75]. That is, nucleosomal DNA sequences tend to be enriched in GC base pairs.

4. The transcription regulation and nucleosome positioning

4.1. The +1 and −1 nucleosomes and nucleosome-free regions (NFRs) near transcription start sites (TSSs)

Nucleosome positioning is in gene regulation since the DNA packing on the surface of the histone octamer can occlude the binding sites of transcription factors (TFs) on genomic DNA. That is to say, the nucleosome positioning at promoters negatively regulates gene transcription by preventing TFs binding. Typically, nucleosomes are depleted around transcription start sites (TSSs), resulting a nucleosome-free region (NFR) that is flanked by two well-positioned nucleosomes (+1 and −1 nucleosomes). In downstream of the TSS, nucleosomes are equally spaced as a nucleosome array. At 3′ direction gene (transcription termination sites (TTS)), there is also a NFR, called 3′ NFR. Additionally, the poly (dA:dT) sequences are found in the 5′ and 3′ NFRs, where they act as nucleosome-excluding sequence. The characteristics of nucleosome organization are found in multiple species, including yeast, worms, flies and humans. In such an organization, the NFRs are often the TFs binding regions. Transcriptional activation involves several steps in yeast [56]. Firstly, special chemical modifications (acetylation and methylation (H3K4me3)) occur on histones of the −1 and +1 nucleosomes (**Figure 12**). The acetylation marks can be recognized by bromodomain modules of the SAGA histone acetyltransferase complex and Bdf1. SAGA and TFIID then deliver TBP to promoters. Then, the pre-initiation complex (PIC) is mounted.

It is suggested that NFRs at promoters result from a competition between TF and nucleosome binding because that incorporating competition with TFs improves the prediction performance for nucleosome positioning, particularly in promoter regions [76]. Moreover, the mechanism is not restricted to a few promoters, but is the typical configuration along the

http://dx.doi.org/10.5772/intechopen.70935

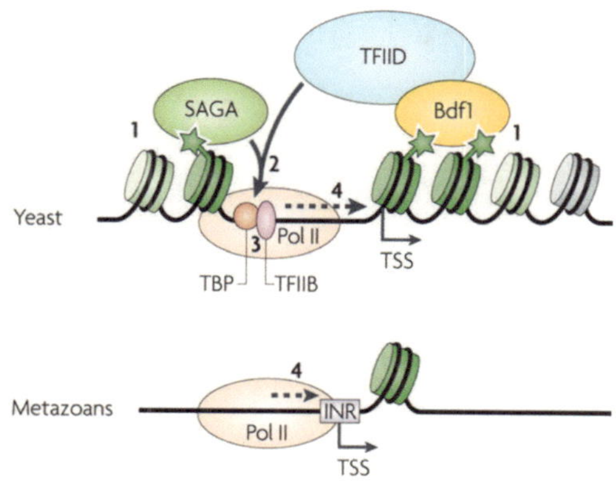

Figure 12. Transcription initiation in budding yeast, including four steps [56].

genomes. Interestingly, it was reported that of the 158 yeast TFs, only 10–20 significantly contribute to inducing NFRs, and these TFs are highly enriched for having direct interactions with chromatin remodelers [76].

Therefore, theoretically, nucleosome level at promoters should negatively associate gene expression level. As expected, for the acid phosphatase inducible PHO5 gene, a significant cell-to-cell variation was found in nucleosome positions and the nucleosome shift correlates with changes of gene expression (**Figure 13**) [77]. However, nucleosome positioning is not absolute, and even with major shifts in gene expression, some cells fail to change nucleosome configuration. We found in human CD4$^+$ T cells, a wider NFR at promoters of housekeeping genes and highly expressed genes [78].

4.2. The difference of nucleosome organization among species

The current studies suggest that almost all eukaryotic organisms hold the nucleosome organization characteristics at the 5′ end of gene, namely a NFR flanked by two (+1 and −1) well-positioned nucleosomes and followed by an array of nucleosomes downstream of TSSs [37, 70, 79]. But compared with multicellular organisms fly, worm, and human, yeast is very simple. Nucleosome organization exhibits some differences. First, averagely, yeast has a short linker DNA. The linker DNA is 18 bp in *S. cerevisiae*, ~28 bp in *Drosophila melanogaster* and *Caenorhabditis elegans*, and ~38 bp in human [56]. Second, the dyad position of the +1 nucleosome relative to TSS appears to vary in different organisms. In yeast, the dyad of the nucleosome is at ~50–60 bp downstream of the TSS [80]. However, in Drosophila, the dyad is found at 135 bp downstream of the TSS, reflecting the differences in transcriptional regulatory mechanisms. Third, the 10–11 bp periodicities of the specific dinucleotides (such as AA/TA/TT-GC, WW-SS) pronounce stronger in yeast than in other multicellular organism. In other words, from single-cell organism to multicellular organism, genomic DNA needs to bear more of "encoding information" to meet a more complex regulation requirement. In genomic DNA of multicellular organism, more of TF binding sites are embedded, which will disturb the coding for other information, such as the coding for nucleosome positioning. Forth, in multicellular organism, there exits exons and introns, thus having a splicing process in transcription. It was

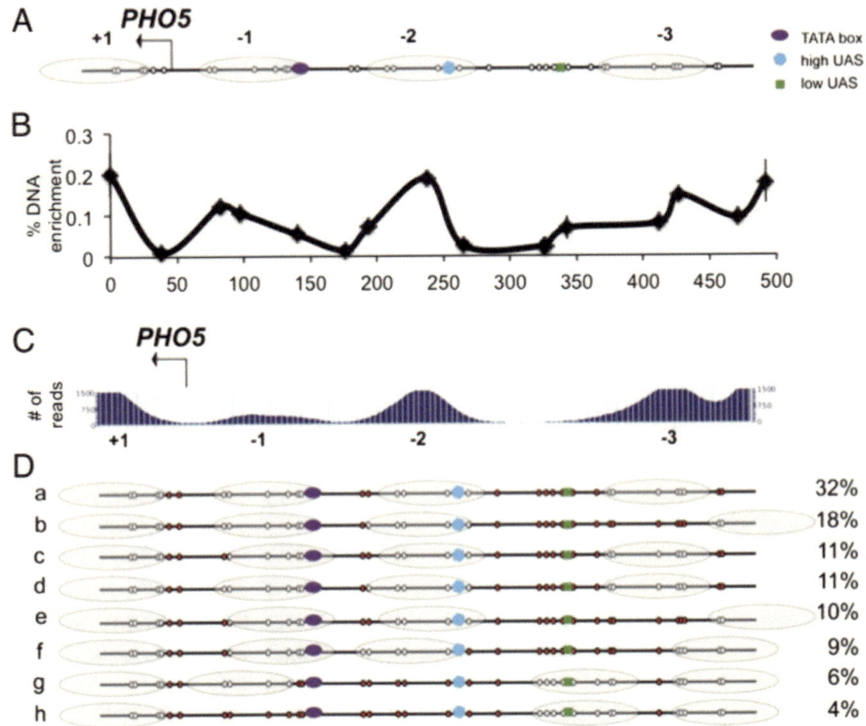

Figure 13. The heterogeneity of nucleosome architecture at the PHO5 promoter. (A) Canonical position of nucleosomes (gray ovals) in the PHO5 promoter. White circles indicate the locations of cytosines of GC dinucleotides. (B) Mapping of mononucleosomal DNA of cells grown in rich media using a nucleosome-scanning assay. Enrichment of mononucleosomal DNA (y axis) is indicated by the midpoints of each amplicon (x axis), and error bars represent 1 SD from two independent biological replicates. (C) MNase-seq track of the PHO5 promoter from cells grown in rich media. (D) Nucleosome architecture of 806 cells from three bulk populations revealed eight conformations (a–h). Nucleosomes are depicted as gray ovals. Red circles indicate methylated cytosines, and white circles indicate unmethylated cytosines that are part of GC dinucleotides. The fraction of total cells that demonstrated each protection pattern is indicated on the right. The SDs for the three experiments were all less than 1.5% [77] (the figure legend is re-written according to the literature).

found that nucleosomes are also well-positioned at both ends of the exon in multicellular [81]. But yeast lacks the feature since its genomic DNA does not include introns. Moreover, upon stress or mutation, nucleosome dynamics frequently occurs at promoters in yeast cells [70]. But in human cells, the nucleosomes alter mainly at enhancers [82, 83].

5. Nucleosome alteration (dynamics) during stress and histone mutation

5.1. Nucleosome alteration upon mutating at modifiable histone residues

Histones are the fundamental element of nucleosomes, and histone mutation do have direct influence on the genome-wide nucleosome organization.

Mutations in histone H3 N-terminal can affect the binding of Chd1, RSC, and SWI/SNF on chromatin, thus having a role in repositioning nucleosomes [84]. Using a native gel electrophoresis experiment, we can quantitively track the loss of nucleosome in different histone mutations. As for influence on RSC repositioning, mutations of H3 R42A and R49A rank the first, both raise the original rate in wild-type nucleosomes up to 2.1-fold. H3 I51A mutations has the least effect on products of RSC directed remodeling (**Figure 14**), indicating that H3 I51A is capable of suppressing the nucleosome-unraveling function of RSC.

SWI/SNF-independent (Sin) mutants have various effects on nucleosome alteration. Class I Sin mutants like H4 R45 has the greatest effect; they may completely evanish certain protein-DNA interactions. Influence of class II mutants is relatively mild, which just do little modification on solvent structure as well as the histone octamer main chain conformation,

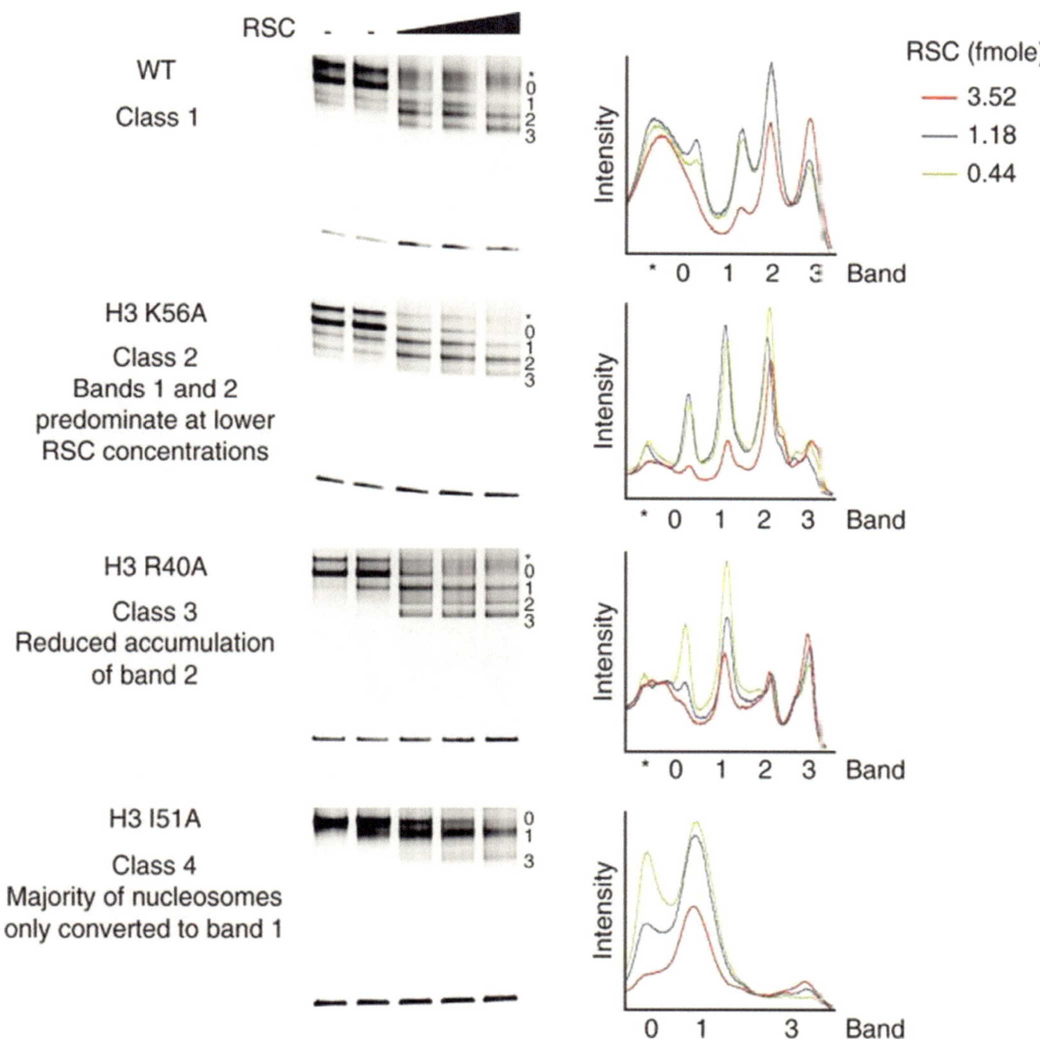

Figure 14. RSC remodeling products in nucleosome with different H3 mutations. H3 mutations could influence the repositioned products, and these mutations are separated into four classes according to the distribution pattern of the products (i.e., bands 1,2 and 3, and band 0 is the origin position). We use four line chart describing band intensities of the RSC remodeled nucleosomes one to one [84] (the legend is rewritten according to the literature).

and class III mutants merely weaken the interactions between octamer and DNA. The changes of protein-NDA interactions lead to an increment of nucleosome-sliding rates.

Histone depletion also has influences on nucleosome. For instance, +1 nucleosomes will notably shift away from the TSS (transcription start sites) when conducting histone H4 depletion in nucleosomes, and +2, +3, and +4 nucleosomes also showed different levels of movement away from the TSS [85]. This was first founded in the study of Harm van Bakel et al., with an excellent idea of researching nucleosome reposition under promoter-closing condition (**Figure 15**).

H3 depletion also causes changes of nucleosome occupancy in genome-wide manner [86]. Depleting HHT1 and using GAL1 promoter to control HHT2 (HHT1 and HHT2 are H3 coding genes) as HHT2 gene nearly does not express when strains grow in dextrose but not in galactose, histone H3 completely disappeared in *S. cerevisiae* (**Figure 16**). In this way, Andrea J. got four strains with different types and carbon sources. In strains with the histone H3 deleted (3 hours), severe changes in nucleosome organization were observed from normal histone levels strain (3 hours) while two types of strains are quite similar in the start (**Figure 16B**). Upon H3 depletion, weakness appears over the whole nucleosomes. An overall view of whole-genome correlation between nucleosome occupancy profiles of normal wild type and H3 depletion strains exhibit an expected decrease compared with ones of H3 shut-off strains grown in galactose (e.g., H3 not deleted) and normal wild type (**Figure 16**). More clearly, there is an evident nucleosome positioning decrement along with the movement from +1 and +2 nucleosomes to the gene's transcription termination site (TTS).

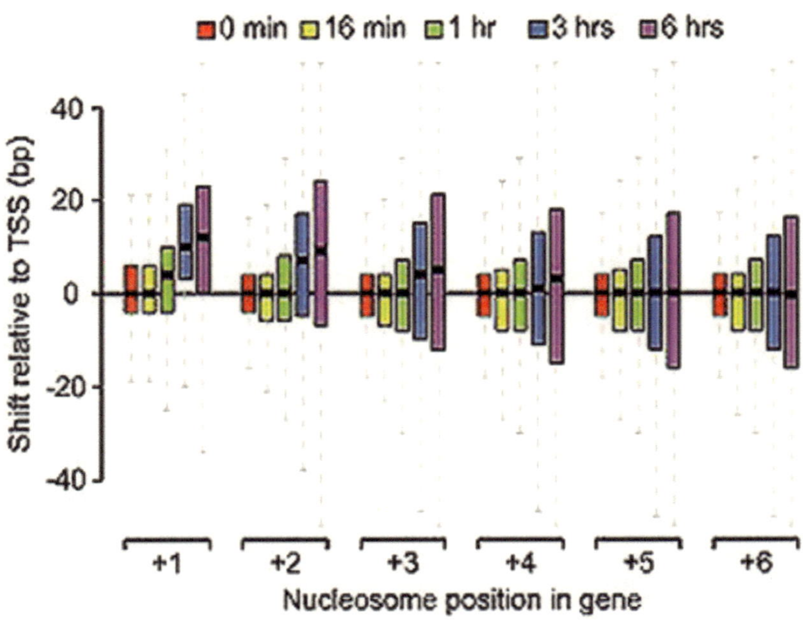

Figure 15. A box plot of nucleosome shifts relative to TSS caused by histone H4 depletion [85].

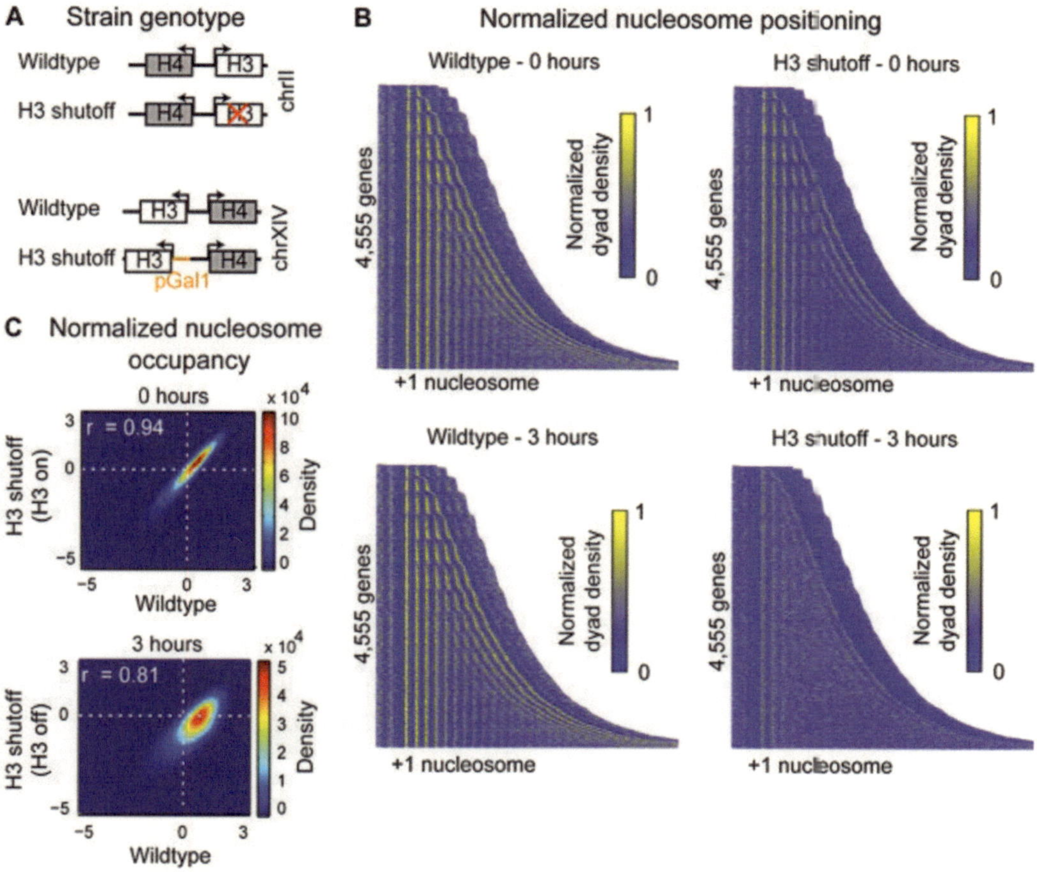

Figure 16. Histone H3 depletion alters nucleosome occupancy genome-wide. (A) Mechanism for H3 depletion in strains. (B) The average, Gaussian-smoothed dyad density for the wild-type and H3 shutoff strains at 0 and 3 hours were aligned on the +1 nucleosomes for 4555 genes. H3 depletion results in a loss of regular positioning of nucleosomes internal to the gene. (C) Log2 normalized average nucleosome occupancy in the wild-type strain versus the H3 shutoff strain. Genome-wide nucleosome occupancy in the two strains is similar prior to H3 depletion (r = 0.94) but decreases following H3 depletion (r = 0.81) [86] (the figure legend is rewritten according to the literature).

5.2. Nucleosome alteration upon heat shock for yeast

Several kinds of changes on carbon source for yeast can alter nucleosome positioning directly or indirectly. After a heat shock, nucleosome occupancy usually becomes higher at promoters that are repressed and the condition is on the contrary at activated ones [87]. A negative correlation is suggested between nucleosome occupancy and transcription levels caused by heat shock. PAPAS is a long non-coding RNA (lncRNA) and was tested carrying out help in the repression of Pol I transcription as it is upregulated by heat shock [88]. CHD4/NuRD is the remodeling complex that could prevent transcription in a way of accessing nucleosomes which should have bound around promoters onto the transcriptional off position. An examination for nucleosome positioning in normal and heat-shocked cells indicated that heat shock led to a promoter-bound nucleosome movement

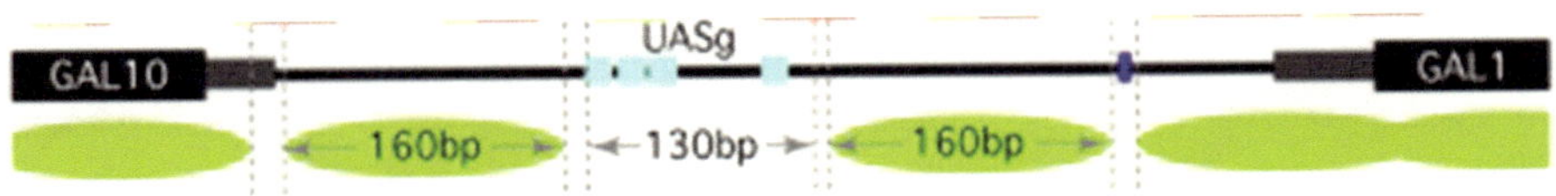

Figure 17. GAL 1, GAL 10, and Gal 4 binding sites (cyan rectangles) locus. Nucleosomes are presented by green circles, and a TATA box is located between Gal4 and GAL1, with blue rectangles [89] (the figure legend is rewritten according to the literature).

award downstream position via promoting PAPAS expression which could induce recruitment of CHD4/NuRD to rDNA [88].

5.3. Nucleosome alteration upon changing carbon source for yeast

Several kinds of changes on carbon source for yeast can alter nucleosome positioning directly or indirectly. Gal4, a transcriptional activator discovered in *S. cerevisiae* has been intensively studied. Two genes, GAL1 and GAL10 are both regulated by Gal4 (**Figure 17**) [89]. It was found that GAL1 promoter nucleosomes became absent from cells grown for many generations in galactose. But by ChIP experiments, Gal4 is found always present both before and after the nutrition shift. In fact, the follow-up Gal80-absence comparison revealed that galactose could remove Gal80 from nucleosomes, an inhibitor of Gal4. Then the recruiting function of freed Gal4 is quickly motivated, leading SWI/SWF binding to the genes. And this always goes with promoter nucleosomes removal as another two ChIP experiments shows.

Besides, the influence of glucose on nucleosome reassembly was affected by the presence of galactose [89]. The transcription factor Msn2, which is recognized with stress-response feature, not only participates in quite a number of environmental stress response as a mediator but also proactively functions in the restructure activities of nucleosome-depleted region (NDR) during transcriptional reprogramming [90]. Msn2 usually binds to small parts of stress response elements (STREs) and a glucose-to-glycerol downshift could apparently promote Msn2 occupancy near STREs (**Figure 18**). Moreover, the nutrition downshift-stress also enables Msn2 to promote the nucleosome repositioning over promoters of genes. It is concluded that Msn2 has a main function of removing the nucleosomes-binding to promoter regions during gene activation and acts negative role in these regions when genes expression is in low level.

5.4. Nucleosome alterations caused by mutations at modifiable histone residues in *S. cerevisiae*

Histone proteins can be modified by chemical modifications on particular residues. We examined the effect of substituting modifiable residues of four core histones with the non-modifiable residue alanine on nucleosome dynamics [70]. We mapped the genome-wide nucleosomes in 22 histone mutants of *S. cerevisiae* and compared the nucleosome alterations relative to the wild-type strain. The results indicated that different types of histone mutation resulted in different phenotypes and a distinct reorganization of nucleosomes. Nucleosome occupancy was altered at telomeres, but not at centromeres. The first nucleosomes upstream (−1) and downstream (+1) of the TSS were more dynamic than other nucleosomes (**Figure 19**). Mutations in histones affected the nucleosome array downstream of the TSS. Highly expressed genes, such as ribosome genes and genes involved in glycolysis, showed increased nucleosome

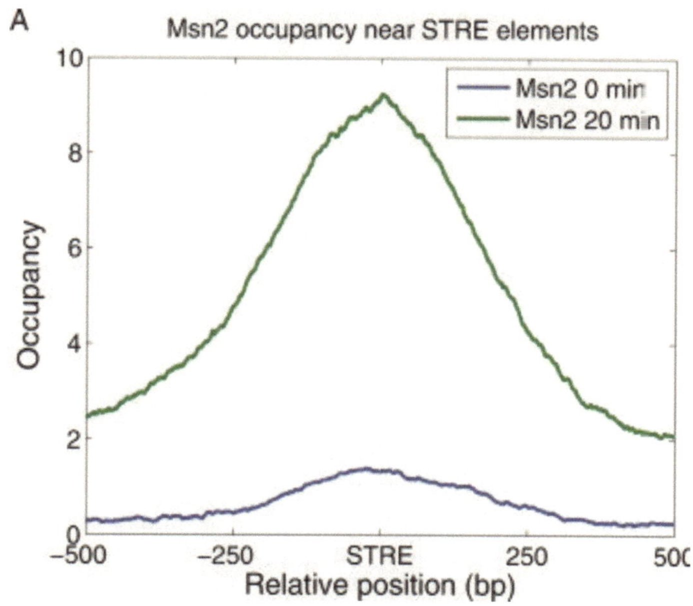

Figure 18. The average Msn2 occupancy near a certain STREs before a carbon source (glucose) replacement with glycerol and 20 min later [89] (the legend is re-written according to the literature).

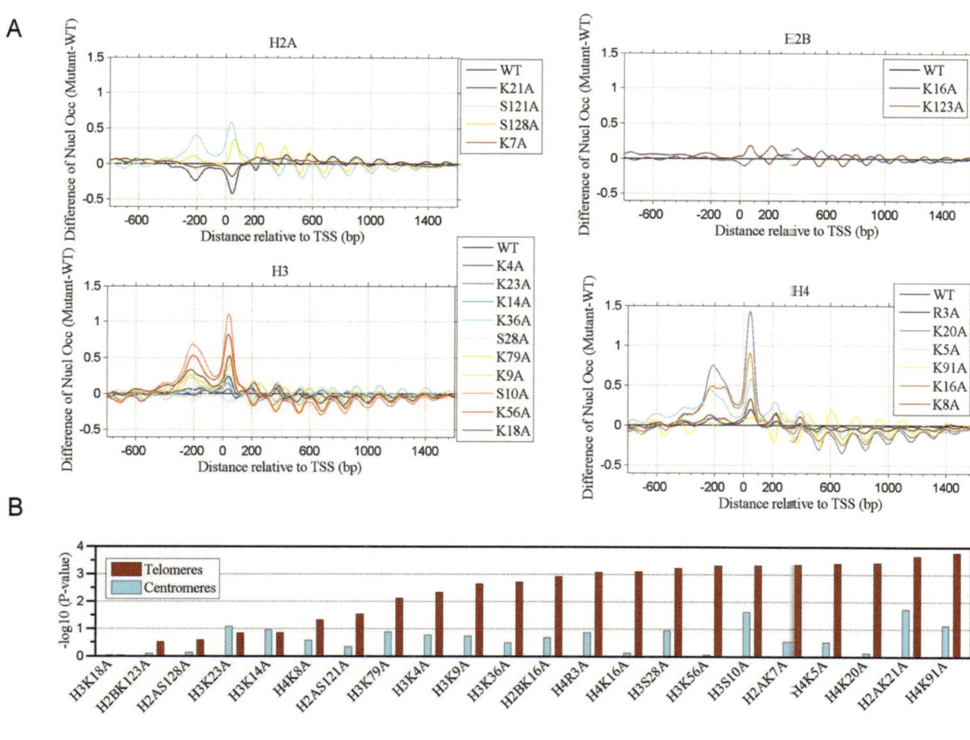

Figure 19. Nucleosome occupancy was altered at telomeres and promoters upon mutating the modifiable residues of the histones. (A) Average differences in nucleosome occupancy around the TSS between mutant and wild-type strains. (B) Difference significance of nucleosome occupancy at telomeres and centromeres between each mutant and the wild-type strain. All P-values (−log10) were calculated with a two-sample t-test [70].

occupancy in many types of histone mutant. In particular, the H3K56A mutant exhibited a high percentage of dynamic genomic regions, decreased nucleosome occupancy at telomeres, increased occupancy at the +1 and −1 nucleosomes, and a slow growth phenotype under stress conditions. Our findings provide insight into the influence of histone mutations on nucleosome dynamics.

6. Htz1 dynamics on chromatin and its effect on nucleosome stability

6.1. Htz1 and nucleosome, Htz1 and transcription

Yeast histone H2A variant Htz1, which is called H2A.Z in mammalian, plays important roles in DNA transactions. Zhang et al. gave a detailed study for the genome-wide dynamics of Htz1 [91]. Firstly, Htz1 occupancy is highly reproducible ($r \geq 0.94$). Secondly, Bdf1 (a component of Swr1 complex), Gcn5 (a histone acetyltransferase) and histone acetylation all play a part in Htz1 occupancy, as well as Swr1. At several specific locations, Swr1 complex is indispensable to meet the requirements for Htz1 deposition. There are obvious correlations between Htz1 and some histone acetylation, implying Htz1 occupies genes in their repressed/basal states, and Htz1 occupancy was reduced in strains with little Gcn5 or Bdf1. Thirdly, Htz1 shows much greater preference than the poor performance of H2A in occupying promoters. Htz1 occupancy is negatively correlated to the presence of a TATA box, suggesting that the occupancy prefers TATA-less promoters. Fourthly, gene activation associates Htz1 loss from promoters.

Zhang et al. presented a model to explain the mechanism that how Htz1 works to regulate transcription (**Figure 20**) [91]. In the particular repressed/basal genes, a nucleosome with Htz1 occupies the promoters and tends to TATA-less regions. Bdf1, a component of SWR1 complex, could promote the process and helps targeting. Loss of Bdf1 could confer a decrement of Htz1 occupancy. SWR1 complex is necessary for deposition as its recruitment involves physical interactions between SWR1 and DNA sequence-specific transcriptional regulators, physical interactions between SWR1 and promoter binding initiation factors and finding histone modification via Bdf1 or other SWR1 components. Gcn5 does not just help target deposition, but also acetylates H3K14 and other residues which may well be the primary reason for the association between Htz1 occupancy and histone acetylation. But Htz1 takes no particular role in making favor of repressing genes, even though it is observed in a high frequency during repression. In fact, Htz1 keeps a balance presence between the repressed and basal states for full activation. When genes states transit from basal to active, chromatin remodeling factors take in action and activators bind to the enhancer. All of the above likely contribute to the Htz1 nucleosome replacement, which promotes activation via giving way to occupancy of certain transcription factors (**Figure 20**).

Martins-Taylor et al. studied Htz1 in a new aspect and revealed that there were some relationship between Htz1 and the cell-cycle progression requirement of establishing transcriptional silencing [92]. Htz1 appeared to work in a direct way to restrict the spread of silent chromatin

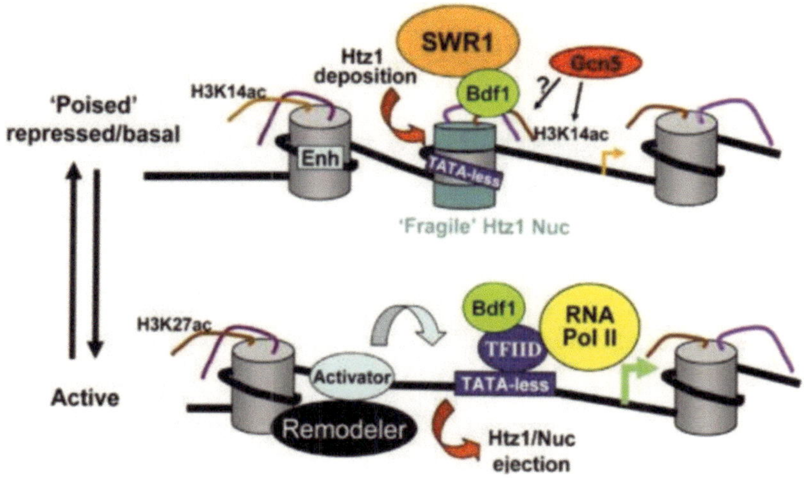

Figure 20. Transcriptional regulation model of Htz1 [91].

from the telomere, and the deletion of genes coding Htz1 could make the establishment of silent chromatin independent from cell-cycle progression.

6.2. Nucleosome, Pol II, Chz1, Htz1, and Spt16

Pol II (RNA polymerase II) promotes the transcription of DNA and is positively associated with the transcription rate. At the beginning of transcription initiation, Pol II are assembled with general transcription factors (GTFs) to make up the pre-initiation complex, binding onto the promoter to initiate transcription [93]. The Htz1 generally occupies the Pol II promoters and affects the combination of GTFs with Pol II, thus inhibiting transcription [91, 94]. Chz1 is an H2B-specific chaperone that delivers Htz1 for H2A substitution [95]. The transcriptional elongation factor FACT is an indispensable component in achieving the process of eliminating the nucleosome block in transcriptional elongation [96]. In yeast cells, Spt16 and Pob3 are the counterparts of FACT. Spt16 destroys the nucleosomes before the running of Pol II complex and reconstructs them after the running. Also, Spt16 has a role of chaperone.

We revealed that Spt16 and Pol II interact with each other and together affect or be affected by gene transcription as they both bind at exposed gene regions, and are positively correlated with the transcription rate (**Figure 21**) [97]. Importantly, Spt16 prefers genes without Htz1 only when Chz1 exists. This discrimination may not be caused for that there are direct interaction mechanism, but is probably to meet the need of transcription initiation. It is found that Chz1 deletion prevents Htz1 occupancy at promoters and telomeres in previous study. Also, in the chz1-deletionmutant, Spt16 binding at ribosomal genes was lost, suggesting that Chz1 is prior in Htz1-bound genes and thus Spt16 has no more binding chances.

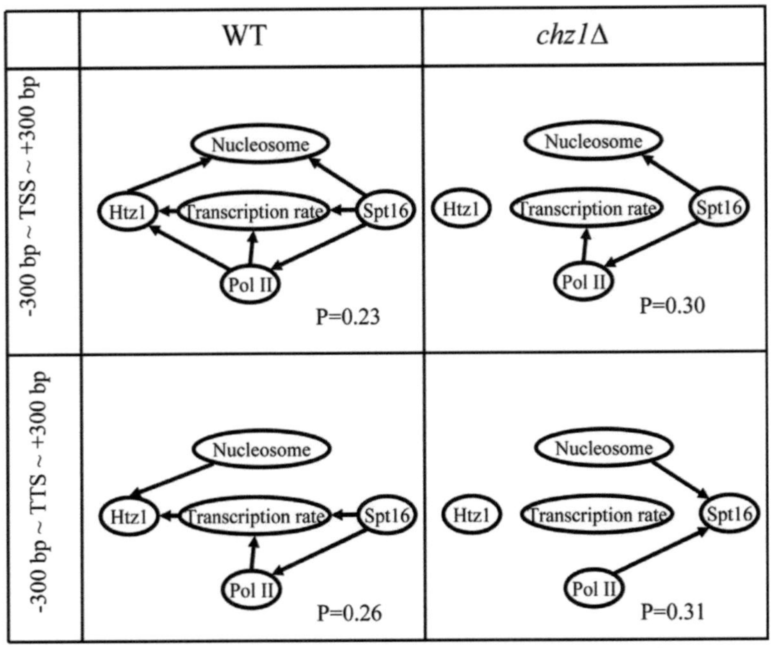

Figure 21. Inferring causal relationships between Spt16, Htz1, Pol II, nucleosome occupancy and transcription rate using Bayesian networks [97]. P-Value indicate probabilities of fitness of the network for test data.

Funding

The writing work was supported by the National Natural Science Foundation of China (No. 31371339 and No. 81660471) and Natural Science Foundation of Xinjiang Province of China (No. 2015211C057).

Author details

Hongde Liu[1]*, Weiheng Ma[1], Jiahao Xie[1], Huamei Li[1], Kun Luo[2], Donghui Luo[2], Lei Liu[3] and Xiao Sun[1]

*Address all correspondence to: liuhongde@seu.edu.cn

1 State Key Laboratory of Bioelectronics, School of Biological Science & Medical Engineering, Southeast University, Nanjing, China

2 Department of Neurosurgery, Xinjiang Evidence-Based Medicine Research Institute, The First Affiliated Hospital of Xinjiang Medical University, Urumqi, China

3 College of Foreign Language & Literature, Northwest Normal University, Lanzhou, China

References

[1] Volpe TA, Kidner C, Hall IM, Teng G, Grewal SIS, Martienssen RA. Regulation of heterochromatic silencing and histone H3 lysine-9 methylation by RNAi. Science. 2002; **297**(5588):1833-1837

[2] Collins FS, Lander ES, Rogers J, Waterston RH, Conso IHGS. Finishing the euchromatic sequence of the human genome. Nature. 2004;**431**(7011):931-945

[3] Luger K, Mader AW, Richmond RK, Sargent DF, Richmond TJ. Crystal structure of the nucleosome core particle at 2.8 angstrom resolution. Nature. 1997;**389**(6648):251-260

[4] Li GH, Zhu P. Structure and organization of chromatin fiber in the nucleus. FEBS Letters. 2015;**589**(20):2893-2904

[5] Arnaudo AM, Molden RC, Garcia BA. Revealing histone variant induced changes via quantitative proteomics. Critical Reviews in Biochemistry and Molecular Biology. 2011;**46**(4):284-294

[6] Reynolds SM, Bilmes JA, Noble WS. Learning a weighted sequence model of the nucleosome core and linker yields more accurate predictions in *Saccharomyces cerevisiae* and *Homo sapiens*. PLoS Computational Biology. 2010;**6**(7):e1000834

[7] Benevolenskaya EV. Histone H3K4 demethylases are essential in development and differentiation. Biochemistry and Cell Biology. 2007;**85**(4):435-443

[8] Barski A, Cuddapah S, Cui K, Roh T-Y, Schones DE, Wang Z, Wei G, Chepelev I, Zhao K. High-resolution profiling of histone methylations in the human genome. Cell. 2007; **129**(4):823-837

[9] Steger DJ, Lefterova MI, Ying L, Stonestrom AJ, Schupp M, Zhuo D, Vakoc AL, Kim JE, Chen JJ, Lazar MA, et al. DOT1L/KMT4 recruitment and H3K79 methylation are ubiquitously coupled with gene transcription in mammalian cells. Molecular and Cellular Biology. 2008;**28**(8):2825-2839

[10] Rosenfeld JA, Wang ZB, Schones DE, Zhao K, DeSalle R, Zhang MQ. Determination of enriched histone modifications in non-genic portions of the human genome. BMC Genomics. 2009;**10**(1):143

[11] Richmond TJ. Hot papers—Crystal structure—Crystal structure of the nucleosome core particle at 2.8 angstrom resolution by K. Luger, A.W. Mader, R.K. Richmond, D.F. Sargent, T.J. Richmond—Comments. Scientist. 1999;**13**(23):15-15

[12] White CL, Suto RK, Luger K. Structure of the yeast nucleosome core particle reveals fundamental changes in internucleosome interactions. The EMBO Journal. 2001;**20**(18): 5207-5218

[13] Yuan GC, Liu YJ, Dion MF, Slack MD, LF W, Altschuler SJ, Rando OJ. Genome-scale identification of nucleosome positions in *S. cerevisiae*. Science. 2005;**309**(5734):626-630

[14] Satchwell SC, Drew HR, Travers AA. Sequence periodicities in chicken nucleosome core DNA. Journal of Molecular Biology. 1986;**191**(4):659-675

[15] Segal E, Fondufe-Mittendorf Y, Chen L, Thastrom A, Field Y, Moore IK, Wang JP, Widom J. A genomic code for nucleosome positioning. Nature. 2006;**442**(7104):772-778

[16] Struhl K, Segal E. Determinants of nucleosome positioning. Nature Structural & Molecular Biology. 2013;**20**(3):267-273

[17] Herzel H, Weiss O, Trifonov EN. 10-11 bp periodicities in complete genomes reflect protein structure and DNA folding. Bioinformatics. 1999;**15**(3):187-193

[18] Zhong J, Luo K, Winter PS, Crawford GE, Iversen ES, Hartemink AJ. Mapping nucleosome positions using DNase-seq. Genome Research. 2016;**26**(3):351-364

[19] Liu H, Wu J, Xie J, Yang X, Lu Z, Sun X. Characteristics of nucleosome core DNA and their applications in predicting nucleosome positions. Biophysical Journal. 2008;**94**(12): 4597-4604

[20] Liu H, Lin S, Cai Z, Sun X. Role of 10-11bp periodicities of eukaryotic DNA sequence in nucleosome positioning. Bio Systems. 2011;**105**(3):295-299

[21] Field Y, Kaplan N, Fondufe-Mittendorf Y, Moore IK, Sharon E, Lubling Y, Widom J, Segal E. Distinct modes of regulation by chromatin encoded through nucleosome positioning signals. PLoS Computational Biology. 2008;**4**(11):e1000216

[22] Kaplan N, Moore IK, Fondufe-Mittendorf Y, Gossett AJ, Tillo D, Field Y, LeProust EM, Hughes TR, Lieb JD, Widom J, et al. The DNA-encoded nucleosome organization of a eukaryotic genome. Nature. 2009;**458**(7236):362-366

[23] Guo SH, Deng EZ, LQ X, Ding H, Lin H, Chen W, Chou KC. iNuc-PseKNC: A sequence-based predictor for predicting nucleosome positioning in genomes with pseudo k-tuple nucleotide composition. Bioinformatics. 2014;**30**(11):1522-1529

[24] Brogaard K, Xi LQ, Wang JP, Widom J. A map of nucleosome positions in yeast at base-pair resolution. Nature. 2012;**486**(7404):496-501

[25] Xi LQ, Fondufe-Mittendorf Y, Xia L, Flatow J, Widom J, Wang JP. Predicting nucleosome positioning using a duration Hidden Markov Model. BMC Bioinformatics. 2010;**11**(1):346

[26] Wang LG, Huang HJ, Dougherty G, Zhao Y, Hossain A, Kocher JPA. Epidaurus: Aggregation and integration analysis of prostate cancer epigenome. Nucleic Acids Research. 2015;**43**(2):e7

[27] Di Gesu V, Lo Bosco G, Pinello L, Yuan GC, Corona DFV. A multi-layer method to study genome-scale positions of nucleosomes. Genomics. 2009;**93**(2):140-145

[28] Locke G, Tolkunov D, Moqtaderi Z, Struhl K, Morozov AV. High-throughput sequencing reveals a simple model of nucleosome energetics. Proceedings of the National Academy of Sciences of the United States of America. 2010;**107**(49):20998-21003

[29] Cui F, Zhurkin VB. Rotational positioning of nucleosomes facilitates selective binding of p53 to response elements associated with cell cycle arrest. Nucleic Acids Research. 2014;**42**(2):836-847

[30] Cui F, Zhurkin VB. Structure-based analysis of DNA sequence patterns guiding nucleosome positioning in vitro. Journal of Biomolecular Structure & Dynamics. 2010;**27**(6):821-841

[31] Levitsky VG, Katokhin AV, Podkolodnaya OA, Furman DP, Kolchanov NA. NPRD: Nucleosome positioning region database. Nucleic Acids Research. 2005;**33**:D67-D70

[32] Nguyen N, Vo A, Won KJ. A wavelet-based method to exploit epigenomic language in the regulatory region. Bioinformatics. 2014;**30**(7):908-914

[33] Stolz RC, Bishop TC. ICM Web: The interactive chromatin modeling web server. Nucleic Acids Research. 2010;**38**:W254-W261

[34] Gabdank I, Barash D, Trifonov EN. FineStr: A web server for single-base-resolution nucleosome positioning. Bioinformatics. 2010;**26**(6):845-846

[35] Chen W, Lin H, Feng PM, Ding C, Zuo YC, Chou KC. iNuc-PhysChem: A sequence-based predictor for identifying nucleosomes via physicochemical properties. PLoS One. 2012;**7**(10):e47843

[36] Weiner A, Hughes A, Yassour M, Rando OJ, Friedman N. High-resolution nucleosome mapping reveals transcription-dependent promoter packaging. Genome Research. 2010;**20**(1):90-100

[37] Chen KF, Xi YX, Pan XW, Li ZY, Kaestner K, Tyler J, Dent S, He XW, Li W. DANPOS: Dynamic analysis of nucleosome position and occupancy by sequencing. Genome Research. 2013;**23**(2):341-351

[38] Meyer CA, He HSH, Brown M, Liu XS. BINOCh: Binding inference from nucleosome occupancy changes. Bioinformatics. 2011;**27**(13):1867-1868

[39] Woo SS, Zhang XK, Sauteraud R, Robert F, Gottardo R. PING 2.0: An R/bioconductor package for nucleosome positioning using next-generation sequencing data. Bioinformatics. 2013;**29**(16):2049-2050

[40] Humburg P, Helliwell CA, Bulger D, Stone G. ChIPseqR: Analysis of ChIP-seq experiments. BMC Bioinformatics. 2011;**12**(1):39

[41] Quintales L, Vazquez E, Antequera F. Comparative analysis of methods for genome-wide nucleosome cartography. Briefings in Bioinformatics. 2015;**16**(4):576-587

[42] Schopflin R, Teif VB, Muller O, Weinberg C, Rippe K, Wedemann G. Modeling nucleosome position distributions from experimental nucleosome positioning maps. Bioinformatics. 2013;**29**(19):2380-2386

[43] Mammana A, Vingron M, Chung HR. Inferring nucleosome positions with their histone mark annotation from ChIP data. Bioinformatics. 2013;**29**(20):2547-2554

[44] Fu K, Tang QZ, Feng JX, Liu XS, Zhang Y. DiNuP: A systematic approach to identify regions of differential nucleosome positioning. Bioinformatics. 2012;**28**(15):1965-1971

[45] Vainshtein Y, Rippe K, Teif VB. NucTools: Analysis of chromatin feature occupancy profiles from high-throughput sequencing data. BMC Genomics. 2017;**18**(1):158

[46] Liu LJ, Xie JM, Sun X, Luo K, Qin ZS, Liu HD. An approach of identifying differential nucleosome regions in multiple samples. BMC Genomics. 2017;**18**(1):135

[47] Lai WKM, Buck MJ. ArchAlign: Coordinate-free chromatin alignment reveals novel architectures. Genome Biology. 2010;**11**:12

[48] Straver R, Oudejans CBM, Sistermans EA, Reinders MJT. Calculating the fetal fraction for noninvasive prenatal testing based on genome-wide nucleosome profiles. Prenatal diagnosis. 2016;**36**(7):614-621

[49] Nellore A, Bobkov K, Howe E, Pankov A, Diaz A, Song JS. NSeq: A multithreaded java application for finding positioned nucleosomes from sequencing data. Frontiers in Genetics. 2012;**3**:320

[50] Lai WK, Bard JE, Buck MJ. ArchTEx: Accurate extraction and visualization of next-generation sequence data. Bioinformatics. 2012;**28**(7):1021-1023

[51] Polishko A, Bunnik EM, Le Roch KG, Lonardi S. PuFFIN—A parameter-free method to build nucleosome maps from paired-end reads. BMC Bioinformatics. 2014;**15**(Suppl 9): S11

[52] Zhang Y, Shin H, Song JS, Lei Y, Liu XS. Identifying positioned nucleosomes with epigenetic marks in human from ChIP-Seq. BMC Genomics. 2008;**9**:537

[53] Vignali M, Hassan AH, Neely KE, Workman JL. ATP-dependent chromatin-remodeling complexes. Molecular and Cellular Biology. 2000;**20**(6):1899-1910

[54] Wang GG, Allis CD, Chi P. Chromatin remodeling and cancer, part II: ATP-dependent chromatin remodeling. Trends in Molecular Medicine. 2007;**13**(9):373-380

[55] Mayes K, Qiu ZJ, Alhazmi A, Landry JW. ATP-dependent chromatin remodeling complexes as novel targets for cancer therapy. Advances in Cancer Research. 2014;**121**:183-233

[56] Jiang CZ, Pugh BF. Nucleosome positioning and gene regulation: Advances through genomics. Nature Reviews Genetics. 2009;**10**(3):161-172

[57] Mavrich TN, Ioshikhes IP, Venters BJ, Jiang C, Tomsho LP, Qi J, Schuster SC, Albert I, Pugh BF. A barrier nucleosome model for statistical positioning of nucleosomes throughout the yeast genome. Genome Research. 2008;**18**(7):1073-1083

[58] Zentner GE, Henikoff S. Surveying the epigenomic landscape, one base at a time. Genome Biology. 2012;**13**(10):250

[59] Meyer CA, Liu XS. Identifying and mitigating bias in next-generation sequencing methods for chromatin biology. Nature Reviews Genetics. 2014;**15**(11):709-721

[60] Tsompana M, Buck MJ. Chromatin accessibility: A window into the genome. Epigenetics & Chromatin. 2014;**7**(1):33

[61] Boyle AP, Davis S, Shulha HP, Meltzer P, Margulies EH, Weng Z, Furey TS, Crawford GE. High-resolution mapping and characterization of open chromatin across the genome. Cell. 2008;**132**(2):311-322

[62] Song L, Crawford GE. DNase-seq: A high-resolution technique for mapping active gene regulatory elements across the genome from mammalian cells. Cold Spring Harbor Protocols. 2010;**2010**(2) pdb. prot5384

[63] Weintraub H, Groudine M. Chromosomal subunits in active genes have an altered conformation. Science. 1976;**193**(4256):848-856

[64] Buenrostro JD, Giresi PG, Zaba LC, Chang HY, Greenleaf WJ. Transposition of native chromatin for fast and sensitive epigenomic profiling of open chromatin, DNA-binding proteins and nucleosome position. Nature Methods. 2013;**10**(12):1213-1218

[65] Buenrostro JD, Wu B, Chang HY, Greenleaf WJ. ATAC-seq: A method for assaying chromatin accessibility genome-wide. Current Protocols in Molecular Biology;**109** 21.29.1-9

[66] Park PJ. ChIP–seq: Advantages and challenges of a maturing technology. Nature Reviews Genetics. 2009;**10**(10):669-680

[67] Giresi PG, Kim J, McDaniell RM, Iyer VR, Lieb JD. FAIRE (formaldehyde-assisted isolation of regulatory elements) isolates active regulatory elements from human chromatin. Genome Research. 2007;**17**(6):877-885

[68] Rhee HS, Pugh BF. ChIP-exo method for identifying genomic location of DNA-binding proteins with near-single-nucleotide accuracy. Current Protocols in Molecular Biology. 2012 Chapter 21:Unit 21.24

[69] Thorvaldsdottir H, Robinson JT, Mesirov JP. Integrative genomics viewer (IGV): High-performance genomics data visualization and exploration. Briefings in Bioinformatics. 2013;**14**(2):178-192

[70] Liu H, Wang P, Liu L, Min Z, Luo K, Wan Y. Nucleosome alterations caused by mutations at modifiable histone residues in *Saccharomyces cerevisiae*. Scientific Reports. 2015; **5**:15583

[71] Feng J, Liu T, Qin B, Zhang Y, Liu XS. Identifying ChIP-seq enrichment using MACS. Nature Protocols. 2012;**7**(9):1728-1740

[72] Benjamini Y, Speed TP. Summarizing and correcting the GC content bias in high-throughput sequencing. Nucleic Acids Research. 2012;**40**(10):e72-e72

[73] Bradnam KR, Seoighe C, Sharp PM, Wolfe KH. G+C content variation along and among *Saccharomyces cerevisiae* chromosomes. Molecular Biology and Evolution. 1999;**16**(5):666-675

[74] Dingwall C, Lomonossoff GP, Laskey RA. High sequence specificity of micrococcal nuclease. Nucleic Acids Research. 1981;**9**(12):2659-2673

[75] Chung HR, Dunkel I, Heise F, Linke C, Krobitsch S, Ehrenhofer-Murray AE, Sperling SR, Vingron M. The effect of micrococcal nuclease digestion on nucleosome positioning data. PLoS One. 2010;**5**(12):e15754

[76] Ozonov EA, van Nimwegen E. Nucleosome free regions in yeast promoters result from competitive binding of transcription factors that interact with chromatin modifiers. PLoS Computational Biology. 2013;**9**(8):e1003181

[77] Small EC, Xi L, Wang JP, Widom J, Licht JD. Single-cell nucleosome mapping reveals the molecular basis of gene expression heterogeneity. Proceedings of the National Academy of Sciences of the United States of America. 2014;**111**(24):E2462-E2471

[78] Liu H, Luo K, Wen H, Ma X, Xie J, Sun X. Quantitative analysis reveals increased histone modifications and a broad nucleosome-free region bound by histone acetylases in highly expressed genes in human CD4+ T cells. Genomics. 2013;**101**(2):113-119

[79] Mavrich TN, Jiang C, Ioshikhes IP, Li X, Venters BJ, Zanton SJ, Tomsho LP, Qi J, Glaser RL, Schuster SC, et al. Nucleosome organization in the drosophila genome. Nature. 2008; **453**(7193):358-362

[80] Radman-Livaja M, Rando OJ. Nucleosome positioning: How is it established, and why does it matter? Developmental Biology. 2010;**339**(2):258-266

[81] Tilgner H, Nikolaou C, Althammer S, Sammeth M, Beato M, Valcarcel J, Guigo R. Nucleosome positioning as a determinant of exon recognition. Nature Structural & Molecular Biology. 2009;**16**(9):996-1001

[82] Skalska L, Stojnic R, Li J, Fischer B, Cerda-Moya G, Sakai H, Tajbakhsh S, Russell S, Adryan B, Bray SJ. Chromatin signatures at Notch-regulated enhancers reveal large-scale changes in H3K56ac upon activation. The EMBO Journal. 2015;**34**(14):1889-1904

[83] Heintzman ND, Hon GC, Hawkins RD, Kheradpour P, Stark A, Harp LF, Ye Z, Lee LK, Stuart RK, Ching CW, et al. Histone modifications at human enhancers reflect global cell-type-specific gene expression. Nature. 2009;**459**(7243):108-112

[84] Somers J, Owen-Hughes T. Mutations to the histone H3 alpha N region selectively alter the outcome of ATP-dependent nucleosome-remodelling reactions. Nucleic Acids Research. 2009;**37**(8):2504-2513

[85] van Bakel H, Tsui K, Gebbia M, Mnaimneh S, Hughes TR, Nislow C. A compendium of nucleosome and transcript profiles reveals determinants of chromatin architecture and transcription. PLoS Genetics. 2013;**9**(5):e1003479

[86] Gossett AJ, Lieb JD. In vivo effects of histone H3 depletion on nucleosome occupancy and position in *Saccharomyces cerevisiae*. PLoS Genetics. 2012;**8**(6):e1002771

[87] Lee CK, Shibata Y, Rao B, Strahl BD, Lieb JD. Evidence for nucleosome depletion at active regulatory regions genome-wide. Nature Genetics. 2004;**36**(8):900-905

[88] Zhao Z, Dammert MA, Hoppe S, Bierhoff H, Grummt I. Heat shock represses rRNA synthesis by inactivation of TIF-IA and lncRNA-dependent changes in nucleosome positioning. Nucleic Acids Research. 2016;**44**(17):8144-8152

[89] Bryant GO, Prabhu V, Floer M, Wang X, Spagna D, Schreiber D, Ptashne M. Activator control of nucleosome occupancy in activation and repression of transcription. PLoS Biology. 2008;**6**(12):2928-2939

[90] Elfving N, Chereji RV, Bharatula V, Bjorklund S, Morozov AV, Broach JR. A dynamic interplay of nucleosome and Msn2 binding regulates kinetics of gene activation and repression following stress. Nucleic Acids Research. 2014;**42**(9):5468-5482

[91] Zhang HY, Roberts DN, Cairns BR. Genome-wide dynamics of Htz1, a histone H2A variant that poises repressed/basal promoters for activation through histone loss. Cell. 2005;**123**(2):219-231

[92] Martins-Taylor K, Sharma U, Rozario T, Holmes SG. H2A.Z (Htz1) controls the cell-cycle-dependent establishment of transcriptional silencing at *Saccharomyces cerevisiae* telomeres. Genetics. 2011;**187**(1):89-104

[93] Sims RJ 3rd, Mandal SS, Reinberg D. Recent highlights of RNA-polymerase-II-mediated transcription. Current Opinion in Cell Biology. 2004;**16**(3):263-271

[94] Albert I, Mavrich TN, Tomsho LP, Qi J, Zanton SJ, Schuster SC, Pugh BF. Translational and rotational settings of H2A.Z nucleosomes across the *Saccharomyces cerevisiae* genome. Nature. 2007;**446**(7135):572-576

[95] Luk E, ND V, Patteson K, Mizuguchi G, WH W, Ranjan A, Backus J, Sen S, Lewis M, Bai Y, et al. Chz1, a nuclear chaperone for histone H2AZ. Molecular Cell. 2007;**25**(3):357-368

[96] Formosa T. FACT and the reorganized nucleosome. Molecular BioSystems. 2008;**4**(11): 1085-1093

[97] Liu H, Luo K, Zhou Z, Mu Y, Wan Y. Histone chaperone Chz1 facilitates the disfavouring property of Spt16 to H2A.Z-containing genes in *Saccharomyces cerevisiae*. The Biochemical Journal. 2014;**460**(3):387-397

Chapter 7

Subunits Common to RNA Polymerases

Cuevas-Bermúdez Abel,
Martínez-Fernández Verónica,
Garrido-Godino Ana I. and Navarro Francisco

Additional information is available at the end of the chapter

http://dx.doi.org/10.5772/intechopen.70936

Abstract

RNA polymerases are heteromultimeric complexes responsible of RNA synthesis. In yeast, as in the other eukaryotes, these complexes contain five common subunits (Rpb5, Rpb6, Rpb8, Rpb10 and Rpb12) that must have similar functions in the three RNA polymerases. However, some of these proteins have been shown to also have specific roles. In the last few decades, substantial progress has been made to understand the role of these common subunits in transcription, but their participation in the activity of each enzyme remains unclear. This review gives a comprehensive overview of current knowledge on the five common subunits of eukaryotic RNA pol, placing attention not only on their common roles in the activity of the RNA pols but also on describing specific roles for some of the complexes.

Keywords: RNA polymerases, transcription, protein complexes, common subunits, RNA

1. Introduction

Transcription is carried out by the RNA polymerases (RNA pol). While archaea and bacteria contain only one RNA pol, most eukarya contain three different enzymes responsible for the specific synthesis of different types of RNAs [1]. RNA pol I synthesises the precursor of the three largest rRNAs, whereas RNA pol III synthesises mostly tRNAs and 5S rRNA, together with several short non-translated RNAs. Meanwhile, RNA pol II produces all mRNAs and many non-coding RNAs [1, 2]. Moreover, in plants, two additional polymerases, IV and V (or nuclear RNA polymerases D and E), reportedly synthesise small interfering RNAs (siRNAs), regulating methylation and participating in gene silencing, as well as long non-coding RNAs involved in development and response to environmental changes [3–5].

While bacteriophage T7 and some related enzymes that transcribe the mitochondrial genome or contribute to chloroplast transcription are single-subunit RNA polymerase [6], bacterial, archaeal and eukaryotic enzymes are heteromultimeric complexes (**Table 1**). As in other eukaryotes, yeast RNA pol I, II and III are composed of 14, 12 and 17 subunits, respectively. These contain a catalytic core formed by the two largest subunits, which are highly conserved through evolution (Rpb1 and Rpb2). Moreover, among all eukaryotic RNA pol subunits, five have bacterial homologues (Rpb1, Rpb2, Rpb3, Rpb6 and Rpb11) and others are common to archaea, but without bacterial homologues (Rpb4, Rpb5, Rpb7, Rpb8, Rpb9, Rpb10, and Rpb12) [1, 2, 6–9]. Finally, eukaryotic RNA pols contain five common subunits to the three enzymes (Rpb5, rpb6, Rpb8, Rpb10 and Rpb12), which have archaeal homologues (**Figure 1**) [10–12].

In the last few decades, substantial progress has been made to understand the role of the RNA pol common subunits in transcription, but their participation in the activity of each enzyme remains unclear. This review gives a comprehensive overview of current knowledge on the

Eukaryotes						
Bacteria	**Archaea**	**RNA pol I**	**RNA pol II**	**RNA pol III**	**RNA pol IV (plants)**	**RNA pol V (plants)**
β	Rpo1 (RpoA)	RPA190	RPB1	RPC160	NRPD1	NRPE1
β	Rpo2 (RpoB)	RPBA135	RPB2	RPC128	NRPD/E2	NRPD/E2
α	Rpo3 (RpoD)	RPAC40	RPB3	RPAC40	RPB3 [1]	RPB3 [1]
α	Rpo11 (RpoL)	RPAC19	RPB11	RPAC19	RPB11	RPB11
ω	Rpo6 (RpoK)	RPB6	RPB6	RPB6	RPB6 [1]	RPB6
	Rpo5 (RpoH)	RPB5	RPB5	RPB5	RPB5 [3]	NRPES5
	Rpb8 (RpoG)*	RPB8	RPB8	RPB8	RPB8 [1]	RPB8 [1]
	Rpo10 (RpoN)	RPB10	RPB10	RPB10	RPB10	RPB10
	Rpo12 (RpoP)	RPB12	RPB12	RPB12	RPB12	RPB12
	Rpo4 (RpoF)	RPA14	RPB4	RPC17	NRPD/E4	NRPD/E4
	Rpo7(RpoE)	RPA43	RPB7	RPC25	NRPD7 [1]	NRPE7
		RPA12	RPB9	RPC11	NRPD9b	RPB9
	Rpo13*					
		RPA49		RPC53		
		RPA34.5		RPC37		
				RPC82		
				RPC34		
				RPC31		

In a square, the RNA pol common subunits in a box. *Subunits RpoG and Rpo13 have been identified only in some archaeal species [6] [1]. The numbers in square brackets indicate the number of orthologues of RNA pol IV and RNA pol V subunits in plants.Different names for common subunits of yeast RNA pol: Rpb5: ABC27; Rpb6: ABC23 or Rpo26; Rpb8: ABC14.5; Rpb10: ABC10β; Rpb12: ABC10α.

Table 1. RNA polymerase (RNA pol) subunit composition.

http://dx.doi.org/10.5772/intechopen.70936

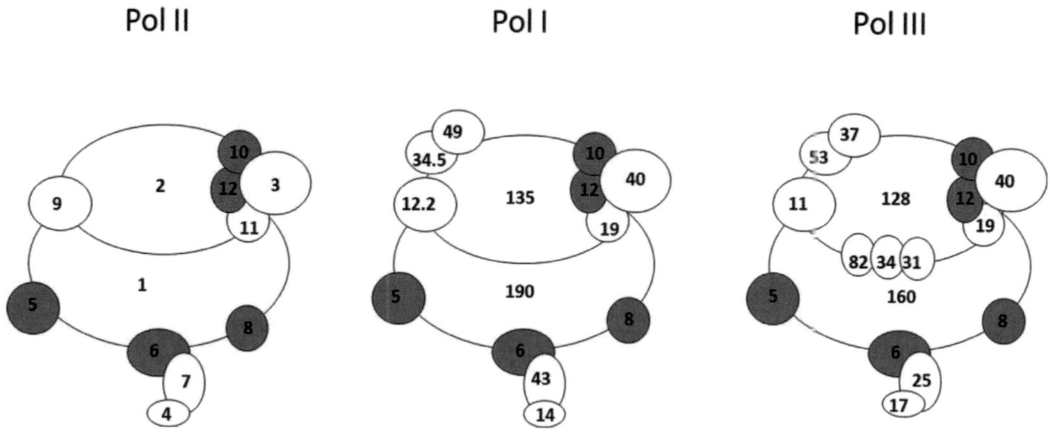

Figure 1. Schematic representation of structure of the RNA pols I, II and III. Each RNA pol common subunit is indicated in grey. The numbers correspond to each subunit are indicated in **Table 1**.

five common subunits of eukaryotic RNA pol, placing attention not only on their common roles in the activity of the RNA pols but also on describing specific roles for some of the complexes.

2. Rpb5

In budding yeast, the essential Rpb5 subunit, also known as ABC27, consists of 215 amino acid residues and has a molecular mass of 27 kDa [11, 13–15]. Contrary to other RNA polymerase common subunits, human Rpb5 homologue (RPB5), with 44% identity and 80% similarity to the yeast polypeptide, fails to complement the *RPB5* null allele in *Saccharomyces cerevisiae* [12]. Rpb5 shows homology not only with a small archaeal subunit called "H" but also with nuclear and cytoplasmic DNA viruses [16, 78]. Rpb5 have two paralogues in *Trypanosome brucei*, *T. cruzi* and *Leishmania major* [17]. Notably, it has been reported that along four distantly related eukaryotic lineages (the higher plant and protistan) Rpb5 shows different isoforms and as a result a diversification of its functions [17].

Structurally, Rpb5 has a bipartite organisation combining two globular modules separated by a short hinge: an N-terminal domain ("*jaw*" domain), found only in eukaryotes (positions 1–142 in *S. cerevisiae*), and a C-terminal globe largely conserved in all non-bacterial enzymes ("*assembly*" domain) [7, 16, 18–20]. Both modules are essential *in vivo* and are functionally exchangeable with their human homologues, except for a small central segment located between positions 121–146 in *S. cerevisiae* [10]. The eukaryotic module of Rpb5 has two highly conserved sequence blocks. One of them harbours the last 12 amino acids and the other, highly conserved (positions 11–30 in budding yeast), belongs to the long hydrophilic helix Rpb5-α1 and occupies the "*lower*" far-end of the DNA *Cleft* [7, 21, 22]. The C-terminal module (position 143–215) binds the largest subunit of RNA pol II (Rpb1) and their paralogues on the RNA pols I and III [10, 23]. Rpb5 does not belong directly to the catalytic domain of RNA pol II [7, 22, 24]. Nevertheless, some studies indicate that the N-terminal domain probably accounts for the Rpb5/DNA contacts found 15–20 nucleotides ahead of the transcription fork in RNA polymerases III [25] and II [26]. In addition, the N-terminal module marks the far end of the DNA channel in the RNA pol II [7, 27] and probably also in the RNA pols I and III [28–30]. Notably, the *lower jaw* and the *assembly* domains of Rpb5 belong to the Shelf module, one of the four RNA pol II mobile modules (*core, jaw-lobe, shelf* and *clamp*) in *S. cerevisiae* [7, 18].

The periphery localization of Rpb5 on all three enzymes [7, 30, 31] would allow possible interactions with general transcription factors or specific gene regulators. It should be the basis of the interaction between Rpb5 and Rsc4, a subunit of RSC (chromatin remodeler complex) in *S. cerevisiae* [32]. The lack of this interaction affects the chromatin structure in the promoter region of some RSC-regulated genes, leading to impaired transcription. Rpb5 also interacts with TFIIE in *Schizosaccharomyces pombe* [33]. In human, RPB5 directly interacts with HBx (hepatitis B virus X protein), essential for HBV infection, and both RPB5 and HBx communicate with transcription initiation factor TFIIB but through different sites [34]. Human RPB5 also interacts with $hTAF_{II}68$ (human TATA-binding protein-associated factor II 68) identified by its homology to the proto-oncogenes EWS (Ewing's sarcoma) and TLS (Translocated in Liposarcoma; another member of the EWS gene family) [35–37]. *In vitro* studies have shown that RPB5 also interacts with the TATA-binding protein-interacting protein 120 (TIP120), which stimulates the transcription driven by RNA pols I and III [38]. Furthermore, RPB5 in human has been described to interact with the general transcription factor TFIIF and this association is critical for the interaction between TFIIF and the RNA pol II [39].

The HBx transactivation seems to be modulated by the protein URI/RMP (Unconventional Prefoldin Rpb5 Interactor) that specifically binds to RPB5 both *in vitro* and *in vivo* and negatively modulates transcription through binding to RPB5 [40]. Owing to RPB5-URI interaction, RPB5 could participate in regulating the androgen receptor in human cells [41]. This interaction also extends to *S. cerevisiae* and the correct association between Rpb5 and the URI orthologue, Bud27, is essential for the correct cytoplasmic assembly of the three RNA pols before their entry to the nucleus [42]. Notably, in mammals, RPB5 forms a complex with UXT, WDR92/Monad, PDRG1, URI, PFDN2 and PFDN6, which is thought to adopt a prefoldin-like structure and cooperates with the cochaperone R2TP complex to assembly of RNA pol II [43–45].

Furthermore, it has been proposed that Bud27 modulates the association between Sth1 (subunit of RSC complex) and the RNA pol II probably through Rpb5 interaction in *S. cerevisiae* [46].

3. Rpb6

Rpb6 (also known as ABC23 or Rpo26) is an acidic 155-amino acid subunit with apparent and predicted molecular masses of 23 and 18 kDa, respectively [11, 47]. It is phosphorylated in all three RNA pols, mainly on serine and threonine residues [48–51]. Moreover, the *in vitro* phosphorylation of rat RPB6 by casein kinase II (CKII) has been demonstrated [52]. Eukaryotic RNA pols I, II and III subunit Rpb6 are homologous in sequence, structure and function to archaeal RNA pol subunit RpoK and bacterial subunit ω [53]. In addition, *S. cerevisiae* Rpb6 is functionally interchangeable with their human homologue *in vivo* [12, 54], demonstrating their structural and functional conservation.

S. cerevisiae RPB6 is an essential gene for cell growth [11, 55], and RNA pol I lacking Rpb6 is virtually inactive in RNA synthesis *in vitro* but regains activity upon the addition of Rpb6 [56].

A role for Rpb6 in transcription elongation has been proposed. In fact, some temperature-sensitive mutants in *S. pombe* are unable to grow in the presence of 6-azauracil and a functional and direct physical interaction of Rpb6 with transcription elongation factor TFIIS has been proposed [57]. Moreover, a recent study demonstrates that the C-terminus of RPAP2, the human homologue of the CTD phosphatase Rtr1 participating in the transition from transcription initiation to elongation, interacts directly with the RNA pol II subunit Rpb6 *in vitro* [58]. Rpb6 could also participate in transcription initiation, since the archaeal TFIIB and Rpb6 counterparts have been demonstrated to interact *in vitro* [59].

According to a global role of Rpb6 in transcription, the *RPB6* gene has also been identified as a dosage suppressor of the cold-sensitive phenotype of *tgs1Δ* cells, which lacks of the trimethylguanosine (TMG) caps of small nuclear (sn) RNA in *S. cerevisiae* [60].

Rpb6 was found to make contact with three small RNA pol subunits, Rpb5, Rpb7 and Rpb8, as well as with the foot of the RNA pol II, with its largest subunits Rpb1 and Rpb2 [7, 61]. Similarly, Rpb6 interacts on the crystal structure with the largest subunit of the RNA pol I, Rpa190 [30] and probably with its homologue in the RNA pol III, Rpc160. Notably, the contact between Rpb6 and Rpb7 involves the residue Gln^{100} of Rpb6 and Gly^{66} of Rpb7 on the RNA pol II core and the *rpb6Q100R* mutant leads to Rpb4/7 dissociation at high temperatures [21, 62].

While the C-terminal segment of Rpb6, from amino acids 72 to 155, is well organised on the crystal structure of yeast RNA pol II [7], the N-terminal domain 71-amino acid segment on the RNA pol II structure, as well as the N-terminal 54-amino acid segment on the RNA pol I structure is disordered [24, 28, 30]. Moreover, the segment from amino acids 55 to 71 of Rpb6 on the RNA pol I structure comprises an α-helix that provides additional contacts with Rpa43 and Rpa14 [28, 30]. The N-terminal region of Rpb6 seems to be dispensable for the functions of this subunit, explaining the lack of conservation of this region with its archaeal homologues and the low degree of similarity of the Rpb6 sequence among various eukaryotes [63]. However, a region of 13 amino acids in the C-terminal domain of Rpb6 is highly conserved in eukaryotes and archaea, suggesting an essential function [63]. Rpb6 is connected to the base of a flexible module containing portions of Rpb1 and Rpb2, called the clamp, through a set of five "switches" that control clamp movement [7]. In addition, the association of Rpb6 with Rpb4/Rpb7 dimmer suggest that these two subunits could modulate the clamp movement and may regulate the position of the clamp by signalling through Rpb6 [62].

Rpb6 and its bacterial homologue have been proposed to promote RNA pol II assembly and/or increase RNA pol stability, through specific interactions with the RNA pol II largest subunit, Rpb1, in the case of *S. cerevisiae* [53, 56, 63, 64]. It has been recently reported that mutations in foot conserved domain of Rpb1 cause an integrity defect of the RNA pol II, altering the association between Rpb1 and Rpb6, and the correct association of the dimer Rpb4/7. This assembly alteration causes a transcriptional defect, which affects the amount of enzyme associated with genes and its transcriptional activity [64]. In addition, the partial dissociation of Rpb4/Rpb7 dimmer leads to an increase in mRNA stability by loss of mRNA imprinting [65, 66]. Notably, all these defects are overcome by *RPB6* overexpression and agree with previous data pointing to an important role of Rpb6 in RNA pol II integrity/assembly [47, 63–65].

In *S. cerevisiae,* assembly of the RNA pols occurs in the cytoplasm prior their entry to the nucleus, and Rpb6 and Rpb5 assemble in a process dependent on the prefoldin-like Bud27 [42]. Similarly, cytoplasmic RNA pol I assembly has been previously proposed in human [44]. In accordance with the role of Rpb6 in RNA pols assembly, the lack of Bud27 alters the correct cytoplasmic assembly of Rpb5 and Rpb6 to the three RNA polymerases, leading to a more instable RNA pol II [42]. Intriguingly, of the five shared subunits, both Rpb6 and Rpb5 have two paralogues in *Trypanosome brucei, T. cruzi* and *Leishmania major* [17]. One is identical in domain organisation to the canonical eukaryotic subunit, called RPB6, whereas the other differs in domain organisation, RPB6z. The highly charged N-terminal domain of RPB6 is absent in RPB6z, making it seems similar in structure to the archaeal subunit. Moreover, the trypanosomatid RPB6z subunit also differs from the canonical RPB6 because of a short insertion in the C-terminal domain [17].

4. Rpb8

Rpb8 (also known as ABC14.5) is an essential subunit of 16.5 kDa conserved among eukaryotes and thought to be restricted to them [11, 12, 67]. However, recently, the Rpb8 archaeal orthologues, called G or Rpo8, has been identified in *Sulfolobus acidocaldarius* (18% identity) and other 15 of the 17 Crenarchaea. This protein presumably appeared at an early step in eukaryotic evolution [6, 68]. This Rpo8 subunit (15.1 kDa; 132 residues) is located at peripheral positions, similar to eukaryotic Rpb8, and interacts with subunit Rpo1N, equivalent to the interaction of Rpb8 with Rpb1 in eukaryotes [69].

Rpb8 crystal structure in RNA pol II contains nine closely packed β-strands forming a double OB-fold [7]. Rpb8 interacts with the largest subunit of the RNA pol II, Rpb1, and shows a subunit interface between Rpb3 and Rpb11. Two-hybrid analyses identified similar binding of Rpb8 to the Rpb1-like subunits of RNA pol I (Rpa190) and RNA pol III (Rpc160) [61]. In addition, mutational analysis of *S. cerevisiae* Rpb8 demonstrated a functional interaction with Rpb6 [67].

As opposed to the Rpb8 human orthologue, *S. pombe* Rpb8 cannot replace *S. cerevisiae* protein. A region of 21 amino acids (residues 68–88) of Rpb8 is absent in *S. pombe.* On the contrary, in human, only six of those residues are missing from the sequence. However, overexpression of Rpc160 in *S. cerevisiae* allows *S. pombe* Rpb8 to functionally replace Rpb8, suggesting a specific interaction between the *S. cerevisiae* Rpb8 and Rpc160 subunit [70]. Notably, *S. pombe* Rpb8 selectively affects RNA polymerase III but not RNA polymerase I complex assembly [70].

5. Rpb10

Rpb10, also called AB10β in yeast, is one of the smallest polypeptides (70-aminoacid polypeptide in *S. cerevisiae* and 71 in *S. pombe*) shared by all three RNA polymerases with a molecular weight of around 10 kDa [71, 72]. Rpb10 has a strong conservation along eukaryotic sequences with 41 identical amino acid positions in fungal, plant and human sequences [73]. In addition, Rpb10 shows a close homology to the N subunit of archaeal enzyme [12, 54, 74] and is loosely related to the smallest enzyme of cytoplasmic DNA viruses [73, 75, 76]. *In vivo* studies in budding yeast have demonstrated that Rpb10 can be functionally replaced by its human homologue (RPB10) [12]. Nevertheless, the N subunit of archaeal cannot replace Rpb10 *in vivo* [73]. However, yeast/archaeal chimeras are largely interchangeable, pointing to a conserved function in their respective transcription complexes [12].

All the eukaryotic forms of Rpb10 share an invariant HVDLIEK motif (located between positions His-53 and Pro-65 in *S. cerevisiae*) critical for the biological activity of Rpb10 [73]. The Rpb10 sequence also harbours an atypical and invariant metal-binding domain $CX_2C...CC$ with Zn^{2+} binding properties *in vitro* [71, 73] that is conserved in eukaryotic, archaeal and viral polypeptides and that is strictly essential for yeast growth, as shown in site-directed mutagenesis experiments [73]. Curiously, mutations out of the metal-chelating domain sequence are fairly tolerant to amino acid replacements [73].

Rpb10 is localised in the periphery of all three RNA polymerases [7, 30, 31]. In budding yeast, Rpb10 was described to interact not only with two essential subunits of the RNA pols I and III, Rpac40 and Rpac19 (homologous to Rpb3 and Rpb11, respectively, in the RNA pol II) but also with the two largest subunits of RNA pol I (Rpa190 and Rpa135) and their homologues in RNA pol III (Rpc160 and Rpc128) [23, 72].

Rpb10 has been found to be involved in the assembly of RNA polymerases in eukaryotes as part of the assembly platform. In fact, it has been proposed that Rpb10 and Rpb12 form a stable complex with Rpb3-Rpb11 (homologous to the bacterial α-subunit homodimer) [77]. Rpb10 and Rpb12 fill concave depressions of Rpb2 and thereby act as structural adaptors between Rpb2 and Rpb3 (reviewed in [1]). Notably, mutations of the invariant HVDLIEK motif lead to a complete depletion of the largest RNA pol I subunit (Rpa190) and decrease the accumulation of mature rRNA species transcribed by RNA pol I [73]. However, Rpb10 could have additional functions beyond RNA polymerases assembly. In accordance, Rpb10 is localised in proximity to TBP in the structural model of the DNA-TBP-TFIIB-RNA pol II transcription initiation complex [79].

6. Rpb12

The eukaryotic subunit Rpb12, also designated as ABC10α [71, 80], together with Rpb10 are the smallest common subunits to the RNA pols. The corresponding gene is essential for growth in *S. cerevisiae* and the lethal phenotype of a yeast *RPB12* null mutant is complemented by expression of its homologous counterparts from *S. pombe* and *Homo sapiens* [12, 81]. A zinc-ribbon motif is conserved in this subunit between eukaryotes and *archaea*. The equivalent to Rpb12 in archaea is the P subunit (RpoP) that shows sequence similarities in their N-terminal zinc ribbon and some highly conserved residues in the C-terminus and that can complement a null RPB12 mutant strain. Mutational analysis of Rpb12 showed that only the first cysteine in the zinc-ribbon motif was essential for viability, whereas the mutation of other three cysteine residues resulted in temperature-sensitive strains [80]. In the crystal structure of RNA pol II from yeast, Rpb12 contacts subunits Rpb2 and Rpb3 [7].

The importance of Rpb12 in transcription is extrapolated from studies on the archaeal P subunit. The P subunit is involved in promoter opening. The ΔP enzyme is unable to form stable open complexes and its activity can be rescued by the addition of Rpb12 or subunit P to transcription reactions. Notably, mutation of cysteine residues in the zinc ribbon impairs the activity of the enzyme in transcription reactions. The conserved zinc ribbon in the N-terminus seems to be important for proper interaction of the complete subunit with other RNA polymerase subunits, and a 17-amino acid C-terminal peptide is sufficient to support all basic RNA polymerase functions *in vitro* [82].

The contact between *S. cerevisiae* RNA pol III and the assembly factor TFIIIC involves the common subunit Rpb12 and the TFIIIB-assembling subunit of TFIIIC, t131. Moreover, thermosensitive mutation in the conserved C-terminal region Rpb12, which weakens this interaction, can be recovered by overexpression of a variant form of t131 [83].

Acknowledgements

This work has been supported by grants from the Spanish Ministry of Economy and Competitiveness, MINECO, and FEDER funds (BFU2013-48643-C3-2-P and BFU2016-77728-C3-2-P to F.N.) and Junta de Andalucía (BIO258).

A.C.B. was a recipient of a FPI predoctoral contract from MINECO. V.M.F. was recipient of a fellowship from Junta de Andalucía and a postdoctoral fellowship from the Junta de Andalucía-University of Jaén. A.I.G-G was a recipient of MEC and a postdoctoral fellowship from the University of Jaén.

Author details

Cuevas-Bermúdez Abel, Martínez-Fernández Verónica, Garrido-Godino Ana I. and Navarro Francisco*

*Address all correspondence to: fngomez@ujaen.es

Departamento de Biología Experimental, Facultad de Ciencias Experimentales, Universidad de Jaén, Paraje de las Lagunillas, Jaén, Spain

References

[1] Werner F, Grohmann D. Evolution of multisubunit RNA polymerases in the three domains of life. Nature Reviews. Microbiology. 2011;**9**:85-98

[2] Werner M, Thuriaux P, Soutourina J. Structure-function analysis of RNA polymerases I and III. Current Opinion in Structural Biology. 2009;**19**:740-745

[3] Haag JR, Pikaard CS. Multisubunit RNA polymerases IV and V: Purveyors of non-coding RNA for plant gene silencing. Nature Reviews. Molecular Cell Biology. 2011; **12**:483-492

[4] Bohmdorfer G, Rowley MJ, Kucinski J, Zhu Y, Amies I, Wierzbicki AT. RNA-directed DNA methylation requires stepwise binding of silencing factors to long non-coding RNA. The Plant Journal. 2014;**79**:181-191

[5] Haag JR, Brower-Toland B, Krieger EK, Sidorenko L, Nicora CD, Norbeck AD, Irsigler A, LaRue H, Brzeski J, McGinnis K, Ivashuta S, Pasa-Tolic L, Chandler VL, Pikaard CS. Functional diversification of maize RNA polymerase IV and V subtypes via alternative catalytic subunits. Cell Rep. 2014;**9**:1-13

[6] Kwapisz M, Beckouet F, Thuriaux P. Early evolution of eukaryotic DNA-dependent RNA polymerases. Trends Genet. 2008

[7] Cramer P, Bushnell DA, Kornberg RD. Structural basis of transcription: RNA polymerase II at 2.8 angstrom resolution. Science. 2001;**292**:1863-1876

[8] Ruprich-Robert G, Thuriaux P. Non-canonical DNA transcription enzymes and the conservation of two-barrel RNA polymerases. Nucleic Acids Research. 2010;**38**:4559-4569

[9] Cramer P, Armache KJ, Baumli S, Benkert S, Brueckner F, Buchen C, Damsma GE, Dengl S, Geiger SR, Jasiak AJ, Jawhari A, Jennebach S, Kamenski T, Kettenberger H, Kuhn CD, Lehmann E, Leike K, Sydow JF, Vannini A. Structure of eukaryotic RNA polymerases. Annual Review of Biophysics. 2008;**37**:337-352

[10] Zaros C, Briand JF, Boulard Y, Labarre-Mariotte S, Garcia-Lopez MC, Thuriaux P, Navarro F. Functional organization of the Rpb5 subunit shared by the three yeast RNA polymerases. Nucleic Acids Research. 2007;**35**:634-647

[11] Woychik NA, Liao SM, Kolodziej PA, Young RA. Subunits shared by eukaryotic nuclear RNA polymerases. Genes & Development. 1990;**4**:313-323

[12] Shpakovski GV, Acker J, Wintzerith M, Lacroix JF, Thuriaux P, Vigneron M. Four subunits that are shared by the three classes of RNA polymerase are functionally interchangeable between *Homo sapiens* and *Saccharomyces cerevisiae*. Molecular and Cellular Biology. 1995;**15**:4702-4710

[13] Navarro F, Thuriaux P. In vivo misreading by tRNA overdose. RNA. 2000;**6**:103-110

[14] Odawara J, Harada A, Yoshimi T, Maehara K, Tachibana T, Okada S, Akashi K, Ohkawa Y. The classification of mRNA expression levels by the phosphorylation state of RNAPII CTD based on a combined genome-wide approach. BMC Genomics. 2011;**12**:516

[15] Nonet M, Scafe C, Sexton J, Young R, Eukaryotic RNA. polymerase conditional mutant that rapidly ceases mRNA synthesis. Molecular and Cellular Biology. 1987;**7**:1602-1611

[16] Langer D, Hain J, Thuriaux P, Zillig W. Transcription in archaea: Similarity to that in eukarya. Proceedings of the National Academy of Sciences of the United States of America. 1995;**92**:5768-5772

[17] Devaux S, Kelly S, Lecordier L, Wickstead B, Perez-Morga D, Pays E, Vanhamme L, Gull K. Diversification of function by different isoforms of conventionally shared RNA polymerase subunits. Molecular Biology of the Cell. 2007;**18**:1293-1301

[18] Cramer P, Bushnell DA, Fu J, Gnatt AL, Maier-Davis B, Thompson NE, Burgess RR, Edwards AM, David PR, Kornberg RD. Architecture of RNA polymerase II and implications for the transcription mechanism. Science. 2000;**288**:640-649

[19] Todone F, Weinzierl RO, Brick P, Onesti S. Crystal structure of RPB5, a universal eukaryotic RNA polymerase subunit and transcription factor interaction target. Proceedings of the National Academy of Sciences of the United States of America. 2000;**97**:6306-6310

[20] Siaut M, Zaros C, Levivier E, Ferri ML, Court M, Werner M, Callebaut I, Thuriaux P, Sentenac A, Conesa C. An Rpb4/Rpb7-like complex in yeast RNA polymerase III contains the orthologue of mammalian CGRP-RCP. Molecular and Cellular Biology. 2003;**23**:195-205

[21] Armache KJ, Mitterweger S, Meinhart A, Cramer P. Structures of complete RNA polymerase II and its subcomplex, Rpb4/7. The Journal of Biological Chemistry. 2005;**280**:7131-7134

[22] Bushnell DA, Kornberg RD. Complete, 12-subunit RNA polymerase II at 4.1-A resolution: Implications for the initiation of transcription. Proceedings of the National Academy of Sciences of the United States of America. 2003;**100**:6969-6973

[23] Flores A, Briand JF, Gadal O, Andrau JC, Rubbi L, Van Mullem V, Boschiero C, Goussot M, Marck C, Carles C, Thuriaux P, Sentenac A, Werner M. A protein-protein interaction map of yeast RNA polymerase III. Proceedings of the National Academy of Sciences of the United States of America. 1999;**96**:7815-7820

[24] Gnatt AL, Cramer P, Fu J, Bushnell DA, Kornberg RD. Structural basis of transcription: An RNA polymerase II elongation complex at 3.3 A resolution. Science. 2001;**292**:1876-1882

[25] Bartholomew B, Durkovich D, Kassavetis GA, Geiduschek EP. Orientation and topography of RNA polymerase III in transcription complexes. Molecular and Cellular Biology. 1993;**13**:942-952

[26] Kim TK, Lagrange T, Wang YH, Griffith JD, Reinberg D, Ebright RH. Trajectory of DNA in the RNA polymerase II transcription preinitiation complex. Proceedings of the National Academy of Sciences of the United States of America. 1997;**94**:12268-12273

[27] Bushnell DA, Cramer P, Kornberg RD. Structural basis of transcription: Alpha-amanitin-RNA polymerase II cocrystal at 2.8 A resolution. Proceedings of the National Academy of Sciences of the United States of America. 2002;**99**:1218-1222

[28] Engel C, Sainsbury S, Cheung AC, Kostrewa D, Cramer P. RNA polymerase I structure and transcription regulation. Nature. 2013;**502**:650-655

[29] Fernandez-Tornero C, Bottcher B, Rashid UJ, Muller CW. Analyzing RNA polymerase III by electron cryomicroscopy. RNA Biology. 2011;**8**:760-765

[30] Fernandez-Tornero C, Moreno-Morcillo M, Rashid UJ, Taylor NM, Ruiz FM, Gruene T, Legrand P, Steuerwald U, Muller CW. Crystal structure of the 14-subunit RNA polymerase I. Nature. 2013;**502**:644-649

[31] Hoffmann NA, Jakobi AJ, Moreno-Morcillo M, Glatt S, Kosinski J, Hagen WJ, Sachse C, Müller CW. Molecular structures of unbound and transcribing RNA polymerase III. Nature. 2015;**528**:231-236

[32] Soutourina J, Bordas-Le Floch V, Gendrel G, Flores A, Ducrot C, Dumay-Odelot H, Soularue P, Navarro F, Cairns BR, Lefebvre O, Werner M. Rsc4 connects the chromatin remodeler RSC to RNA polymerases. Molecular and Cellular Biology. 2006;**26**:4920-4933

[33] Hayashi K, Watanabe T, Tanaka A, Furumoto T, Sato-Tsuchiya C, Kimura M, Yokoi M, Ishihama A, Hanaoka F, Ohkuma Y. Studies of *Schizosaccharomyces pombe* TFIIE indicate conformational and functional changes in RNA polymerase II at transcription initiation. Genes to Cells. 2005;**10**:207-224

[34] Lin Y, Nomura T, Cheong J, Dorjsuren D, Iida K, Murakami S. Hepatitis B virus X protein is a transcriptional modulator that communicates with transcription factor IIB and the RNA polymerase II subunit 5. The Journal of Biological Chemistry. 1997;**272**:7132-7139

[35] Bertolotti A, Lutz Y, Heard DJ, Chambon P, Tora L. hTAF (II) 68, a novel RNA/ssDNA-binding protein with homology to the pro-oncoproteins TLS/FUS and EWS is associated with both TFIID and RNA polymerase II. The EMBO Journal. 1996;**15**:5022

[36] Bertolotti A, Melot T, Acker J, Vigneron M, Delattre O, Tora L. EWS, but not EWS-FLI-1, is associated with both TFIID and RNA polymerase II: Interactions between two members of the TET family, EWS and hTAFII68, and subunits of TFIID and RNA polymerase II complexes. Molecular and Cellular Biology. 1998;**18**:1489-1497

[37] Morohoshi F, Arai K, Takahashi E-i, Tanigami A, Ohki M. Cloning and mapping of a human RBP56 gene encoding a putative RNA binding protein similar to FUS/TLS and EWS proteins. Genomics. 1996;**38**:51-57

[38] Makino Y, Yogosawa S, Kayukawa K, Coin F, Egly JM, Wang Z, Roeder RG, Yamamoto K, Muramatsu M, Tamura T. TATA-binding protein-interacting protein 120, TIP120, stimulates three classes of eukaryotic transcription via a unique mechanism. Molecular and Cellular Biology. 1999;**19**:7951-7960

[39] Le TT, Zhang S, Hayashi N, Yasukawa M, Delgermaa L, Murakami S. Mutational analysis of human RNA polymerase II subunit 5 (RPB5): The residues critical for interactions with TFIIF subunit RAP30 and hepatitis B virus X protein. Journal of Biochemistry (Tokyo). 2005;**138**:215-224

[40] Dorjsuren D, Lin Y, Wei W, Yamashita T, Nomura T, Hayashi N, Murakami S. RMP, a novel RNA polymerase II subunit 5-interacting protein, counteracts transactivation by hepatitis B virus X protein. Molecular and Cellular Biology. 1998;**18** 7546-7555

[41] P. Mita, J.N. Savas, N. Djouder, J.R. Yates 3rd, S. Ha, R. Ruoff, E.D. Schafler, J.C. Nwachukwu, N. Tanese, N.J. Cowan, J. Zavadil, M.J. Garabedian, S.K. Logan, Regulation

of androgen receptor mediated transcription by Rpb5 binding protein URI/RMP, Molecular and Cellular Biology. 2011;**31**:3639-3652

[42] Miron-Garcia MC, Garrido-Godino AI, Garcia-Molinero V, Hernandez-Torres F, Rodriguez-Navarro S, Navarro F. The prefoldin bud27 mediates the assembly of the eukaryotic RNA polymerases in an rpb5-dependent manner. PLoS Genetics. 2013;**9**:e1003297

[43] Gstaiger M, Luke B, Hess D, Oakeley EJ, Wirbelauer C, Blondel M, Vigneron M, Peter M, Krek W. Control of nutrient-sensitive transcription programs by the unconventional prefoldin URI. Science. 2003;**302**:1208-1212

[44] Boulon S, Pradet-Balade B, Verheggen C, Molle D, Boireau S, Georgieva M, Azzag K, Robert MC, Ahmad Y, Neel H, Lamond AI, Bertrand E. HSP90 and its R2TP/prefoldin-like cochaperone are involved in the cytoplasmic assembly of RNA polymerase II. Molecular Cell. 2010;**39**:912-924

[45] Boulon S, Bertrand E, Pradet-Balade B. HSP90 and the R2TP co-chaperone complex: Building multi-protein machineries essential for cell growth and gene expression. RNA Biology. 2012;**9**:1-8

[46] Miron-Garcia MC, Garrido-Godino AI, Martinez-Fernandez V, Fernandez-Pevida A, Cuevas-Bermudez A, Martin-Exposito M, Chavez S, de la Cruz J, Navarro F. The yeast prefoldin-like URI-orthologue Bud27 associates with the RSC nucleosome remodeler and modulates transcription. Nucleic Acids Research. 2014;**42**:9666-9676

[47] Archambault J, Schappert KT, Friesen JD. A suppressor of an RNA polymerase II mutation of *Saccharomyces cerevisiae* encodes a subunit common to RNA polymerases I, II, and III. Molecular and Cellular Biology. 1990;**10**:6123-6131

[48] Bell GI, Valenzuela P, Rutter WJ. Phosphorylation of yeast RNA polymerases. Nature. 1976;**261**:429-431

[49] Bell GI, Valenzuela P, Rutter WJ. Phosphorylation of yeast DNA-dependent RNA polymerases in vivo and in vitro. Isolation of enzymes and identification of phosphorylated subunits. The Journal of Biological Chemistry. 1977;**252**:3082-3091

[50] Buhler JM, Iborra F, Sentenac A, Fromageot P. The presence of phosphorylated subunits in yeast RNA polymerases A and B. FEBS Letters. 1976;**72**:37-41

[51] Kolodziej PA, Woychik N, Liao SM, Young RA. RNA polymerase II subunit composition, stoichiometry, and phosphorylation. Molecular and Cellular Biology. 1990;**10**:1915-1920

[52] Kayukawa K, Makino Y, Yogosawa S, Tamura T. A serine residue in the N-terminal acidic region of rat RPB6, one of the common subunits of RNA polymerases, is exclusively phosphorylated by casein kinase II in vitro. Gene. 1999;**234**:139-147

[53] Minakhin L, Bhagat S, Brunning A, Campbell EA, Darst SA, Ebright RH, Severinov K. Bacterial RNA polymerase subunit omega and eukaryotic RNA polymerase subunit RPB6 are sequence, structural, and functional homologs and promote RNA polymerase assembly. Proceedings of the National Academy of Sciences of the United States of America. 2001;**98**:892-897

[54] McKune K, Moore PA, Hull MW, Woychik NA. Six human RNA polymerase subunits functionally substitute for their yeast counterparts. Molecular and Cellular Biology. 1995;**15**:6895-6900

[55] McKune K, Woychik NA. Functional substitution of an essential yeast RNA polymerase subunit by a highly conserved mammalian counterpart. Molecular and Cellular Biology. 1994;**14**:4155-4159

[56] Lanzendorfer M, Smid A, Klinger C, Schultz P, Sentenac A, Carles C, Riva M. A shared subunit belongs to the eukaryotic core RNA polymerase. Genes & Development. 1997; **11**:1037-1047

[57] Ishiguro A, Nogi Y, Hisatake K, Muramatsu M, Ishihama A. The Rpb6 subunit of fission yeast RNA polymerase II is a contact target of the transcription elongation factor TFIIS. Molecular and Cellular Biology. 2000;**20**:1263-1270

[58] Wani S, Hirose Y, Ohkuma Y. Human RNA polymerase II-associated protein 2 (RPAP2) interacts directly with the RNA polymerase II subunit Rpb6 and participates in pre-mRNA 3′-end formation. Drug Discoveries & Therapeutics. 2014;**8**:255-261

[59] Magill CP, Jackson SP, Bell SD. Identification of a conserved archaeal RNA polymerase subunit contacted by the basal transcription factor TFB. The Journal of Biological Chemistry. 2001;**276**:46693-46696

[60] Qiu ZR, Schwer B, Shuman S. Two routes to genetic suppression of RNA trimethylguanosine cap deficiency via C-terminal truncation of U1 snRNP subunit Snp1 or overexpression of RNA polymerase subunit Rpo26. G3 (Bethesda). 2015;**5**:1361-1370

[61] Briand JF, Navarro F, Rematier P, Boschiero C, Labarre S, Werner M, Shpakovski GV, Thuriaux P. Partners of Rpb8p, a small subunit shared by yeast RNA polymerases I, II and III. Molecular and Cellular Biology. 2001;**21**:6056-6065

[62] Tan Q, Prysak MH, Woychik NA. Loss of the Rpb4/Rpb7 subcomplex in a mutant form of the Rpb6 subunit shared by RNA polymerases I, II, and III. Molecular and Cellular Biology. 2003;**23**:3329-3338

[63] Nouraini S, Archambault J, Friesen JD. Rpo26p, a subunit common to yeast RNA polymerases, is essential for the assembly of RNA polymerases I and II and for the stability of the largest subunits of these enzymes. Molecular and Cellular Biology. 1996;**16**:5985-5996

[64] Garrido-Godino AI, Garcia-Lopez MC, Navarro F. Correct assembly of RNA polymerase II depends on the foot domain and is required for multiple steps of transcription in *Saccharomyces cerevisiae*. Molecular and Cellular Biology. 2013;**33**:3611-3626

[65] Garrido-Godino AI, Garcia-Lopez MC, Garcia-Martinez J, Pelechano V, Medina DA, Perez-Ortin JE, Navarro F. Rpb1 foot mutations demonstrate a major role of Rpb4 in mRNA stability during stress situations in yeast. Biochimica et Biophysica Acta. 2016; **1859**:731-743

[66] Shalem O, Groisman B, Choder M, Dahan O, Pilpel Y. Transcriptome kinetics is governed by a genome-wide coupling of mRNA production and degradation: A role for RNA Pol II. PLoS Genetics. 2011;**7**:e1002273

[67] Briand JF, Navarro F, Gadal O, Thuriaux P. Cross talk between tRNA and rRNA synthesis in *Saccharomyces cerevisiae*. Molecular and Cellular Biology. 2001;**21**:189-195

[68] Koonin EV, Makarova KS, Elkins JG. Orthologs of the small RPB8 subunit of the eukaryotic RNA polymerases are conserved in hyperthermophilic Crenarchaeota and "Korarchaeota". Biology Direct. 2007;**2**:38

[69] Korkhin Y, Unligil UM, Littlefield O, Nelson PJ, Stuart DI, Sigler PB, Bell SD, Abrescia NG. Evolution of complex RNA polymerases: The complete archaeal RNA polymerase structure. PLoS Biology. 2009;**7**:e1000102

[70] Voutsina A, Riva M, Carles C, Alexandraki D. Sequence divergence of the RNA polymerase shared subunit ABC14.5 (Rpb8) selectively affects RNA polymerase III assembly in *Saccharomyces cerevisiae*. Nucleic Acids Research. 1999;**27**:1047-1055

[71] Carles C, Treich I, Bouet F, Riva M, Sentenac A. Two additional common subunits, ABC10 alpha and ABC10 beta, are shared by yeast RNA polymerases. Journal of Biological Chemistry. 1991;**266**:24092-24096

[72] Lalo D, Carles C, Sentenac A, Thuriaux P. Interactions between three common subunits of yeast RNA polymerases I and III. Proceedings of the National Academy of Sciences of the United States of America. 1993;**90**:5524-5528

[73] Gadal O, Shpakovski GV, Thuriaux P. Mutants in ABC10beta, a conserved subunit shared by all three yeast RNA polymerases, specifically affect RNA polymerase I assembly. The Journal of Biological Chemistry. 1999;**274**:8421-8427

[74] Lanzendörfer M, Langer D, Hain J, Klenk H-P, Holz I, Arnold-Ammer I, Zillig W. Structure and function of the DNA-dependent RNA polymerase of *Sulfolobus*. Systematic and Applied Microbiology. 1993;**16**:656-664

[75] Amegadzie BY, Ahn B-Y, Moss B. Characterization of a 7-kilodalton subunit of vaccinia virus DNA-dependent RNA polymerase with structural similarities to the smallest subunit of eukaryotic RNA polymerase II. Journal of Virology. 1992;**66**:3003-3010

[76] Iyer LM, Balaji S, Koonin EV, Aravind L. Evolutionary genomics of nucleo-cytoplasmic large DNA viruses. Virus Research. 2006;**117**:156-184

[77] Wild T, Cramer P. Biogenesis of multisubunit RNA polymerases. Trends in Biochemical Sciences. 2012;**37**:99-105

[78] Iyer V, Struhl K. Absolute mRNA levels and transcriptional initiation rates in Saccharomyces cerevisiae. Proceedings of the National Academy of Sciences of the United States of America. 1996;**93**:5208-5212

http://dx.doi.org/10.5772/intechopen.70936

[79] Kostrewa D, Zeller ME, Armache KJ, Seizl M, Leike K, Thomm M, Cramer P. RNA polymerase II-TFIIB structure and mechanism of transcription initiation. Nature. 2009;**462**: 323-330

[80] Rubbi L, Labarre-Mariotte S, Chedin S, Thuriaux P. Functional characterization of ABC10alpha, an essential polypeptide shared by all three forms of eukaryotic DNA-dependent RNA polymerases. The Journal of Biological Chemistry. 1999;**274**:31485-31492

[81] Treich I, Carles C, Riva M, Sentenac A. RPC10 encodes a new mini subunit shared by yeast nuclear RNA polymerases. Gene Expression. 1992;**2**:31-37

[82] Reich C, Zeller M, Milkereit P, Hausner W, Cramer P, Tschochner H, Thomm M. The archaeal RNA polymerase subunit P and the eukaryotic polymerase subunit Rpb12 are interchangeable in vivo and in vitro. Molecular Microbiology. 2009;**71**:989-1002

[83] Dumay H, Rubbi L, Sentenac A, Marck C. Interaction between yeast RNA polymerase III and transcription factor TFIIIC via ABC10α and τ131 subunits. Journal of Biological Chemistry. 1999;**274**:33462-33468